GO FOR LAUNCH

An Illustrated History
of Cape Canaveral

by Joel Powell
with Art LeBrun

Dedicated to all the engineers, scientists, technicians, administrators and military personnel who have worked at Cape Canaveral from 1950 to the present day (and especially to the contractor photographers who created most of the photographs featured in this book). Your efforts to defend your country and advance the field of rocketry also helped to open the space frontier, for which we will be forever grateful.

All rights reserved under article two of the Berne Copyright Convention (1971).

Published by Apogee Books, Burlington,
Ontario, Canada, L7R 1Y9 http://www.apogeebooks.com
Tel: 905 637 5737

Printed and bound in Canada
Go For Launch by Joel Powell
©2019 Apogee Books
ISBN 1-926592-13-8
ISBN13 978-1-926592-13-8

Cape Canaveral in Florida as seen from an orbiting Space Shuttle.

CONTENTS

Foreword 7
Cape Canaveral 10

PART I

Missile Test Center Operations	39
Bumper (V-2) (at Launch Complex 3)	40
The Cruise Missiles	
Lark	40
Matador and Mace	41
Snark	43
Bomarc	45
Bull Goose (LC-1,2,3,4; 21 and 22)	46
Navaho (LC-9)	47
Medium Range Missiles	
Redstone (LC-4)	50
Redstone and Jupiter A	52
Pershing I (LC-30)	53
Pershing II (LC-16)	54
Intermediate Range Missiles	
Thor (LC-17)	55
Jupiter IRBM (LC-26)	60
The Navy Missiles	
Polaris FTV (LC-3)	65
Polaris A-1, A-2, A-3 (LC-25, 29)	65
Polaris Shipboard Launches	67
Royal Navy Chevaline (LC-29)	68
Poseidon (LC-25, 29)	68
Trident I (LC-29)	69
Trident II (LC-46)	70
NOTU (Naval Ordnance Test Unit)	71
Sea Launches	72
Intercontinental Ballistic Missiles	
Atlas (LC-11, 12, 13, 14)	74
Titan I (LC-15, 16; 19, 20)	81
Minuteman I, II, III (LC-31/32)	85
Titan II (LC-15, 16)	89
The Research Birds	
Hermes RV-A-10 (LC-3)	93
X-17 (LC-3)	94
Jason (LC-4, 10)	95
Jupiter C (LC-5)	96
Blue Scout Jr. (LC-18A)	97
Blue Scout I, II (LC-18B)	99
Weather Rockets at LC-43 and LC-47	101
Sounding Rockets at Cape Canaveral	102
Aerial Research and Development	
Alpha Draco (LC-4, 10)	103
ALBM (Bold Orion and High Virgo)	105
Bomarc (B-57 target aircraft)	106
Navaho X-10	107
Hound Dog	107
Skybolt	108
A Variety of Launch Facilities	108
The Industrial Area (missile hangars)	115
Mission Control Center	118
The Skid Strip	120
Incidents and Accidents	121
Downrange Operations	138
The Recovery of Liberty Bell 7	142
Air Force Salvage Crews	143
Photographic Services at Cape Canaveral	145

PART II

The Space Launch Era (1958 to 1986)   149

Significant Payloads   150
Juno I (LC-26)   153
Juno II (LC-26; 5/6)   156
Vanguard (LC-18A)   159
Thor-Able (LC-17)   163
Atlas-SCORE (LC-11)   167
Atlas-Able (LC-12, 14)   168
Atlas-Agena (LC-12, 13, 14)   171
Mercury-Redstone (LC-5)   178

Astronauts & Presidents at the Cape   183
Mercury-Atlas (LC-14)   188
Thor-Able-Star (LC-17)   196
Thor-Delta & Delta (LC-17)   199
Atlas-Centaur (LC-36)   206
Titan IIIA (LC-20)   211
Titan IIIC (LC-40/41/ITL)   211
Gemini-Titan (LC-19)   218
Saturn I, IB (LC-34, 37)   223

Research Vehicles
Big Shot/Echo (Thor, LC-17)   228
ASSET (Thor, LC-17)   230
FIRE (Atlas-Antares, LC-12)   233
STAFF (Polaris A-1)   236
Classified Atlas-Agenas at pad 13   237
Titan 34D & Commercial Titan   239

PART III

Space Launch in the Post-Challenger Era   243

Titan IV (LC-40, 41/ITL)   244
Delta II (LC-17)   247
Atlas I, II, III (LC-36)   251
Delta III (LC-17B)   256

Research Rockets
Starbird, Aries, Joust, LCLV (at LC-20)   256
Athena (LC-46)   257
Atlas V (LC-41)   258
Delta IV (LC-37B)   263

Press Coverage at Cape Canaveral   266

The Ruins of Cape Canaveral   268

Monuments and Gate Guardians   275

Appendix:
1980's Tour Guide Reproduction   305

Cape Lore   314

Bibliography   317

Photo credits   318

Index   319

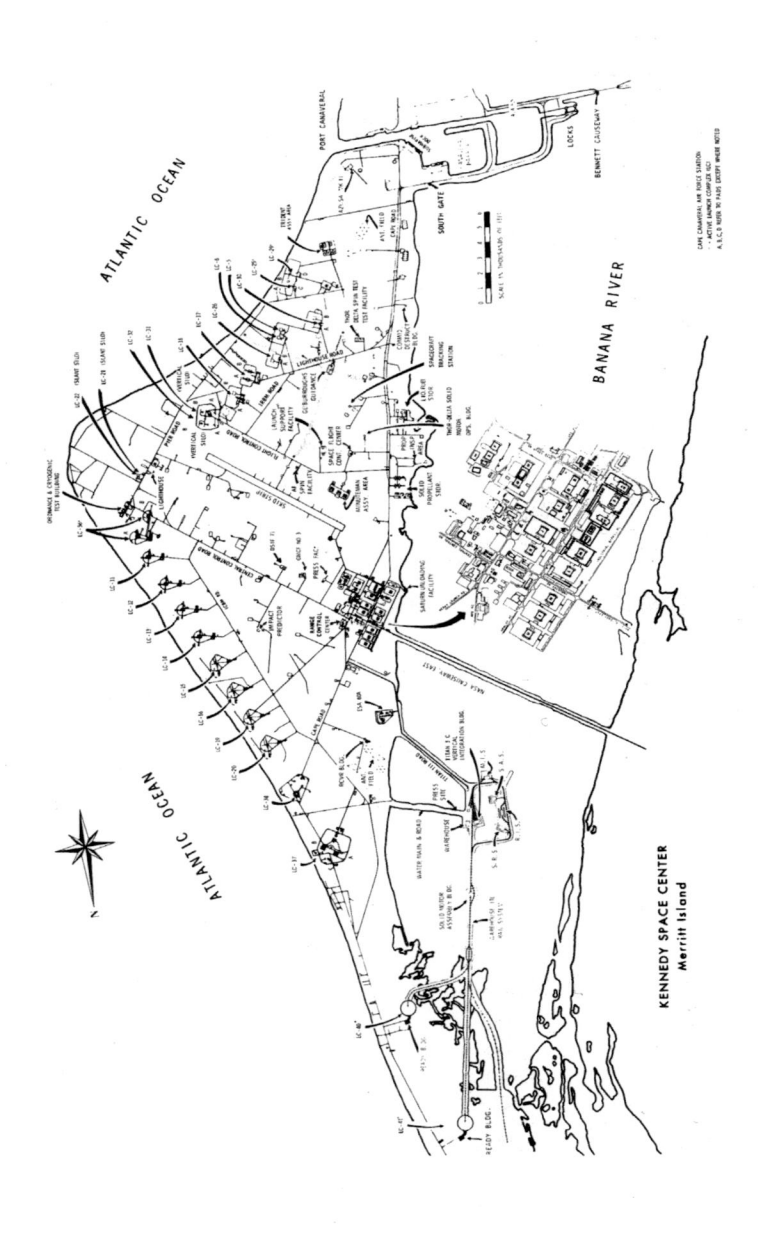

Detailed map of Cape Canaveral circa 1975.

Foreword

Cape Canaveral is mostly quiet now. Where once the roar of a rocket's launch split the air of this seaside military base once every two days on average at the peak of activity in 1960, there are now only three operational launch complexes at Cape Canaveral that support just a handful of launches every year (in 2005 there were just six orbital launches from the Cape!).

With the Cold War now just a fading memory, the weapons test flights of that era are gone now, their abandoned launch pads rusting in the tangy salt air and relentless Florida sunshine. Only the Trident missile launches from submerged nuclear submarines off the Florida coast are still conducted (but are usually invisible from shore). Most of the familiar launch vehicles that began flying in the 1960's (Atlas-Centaur, Titan IV) are also retired now, with the exception of the Delta II that is still operational at Complex 17. Delta II has a nearly flawless record (with but a single, spectacular failure) since it was introduced in 1989. The purpose of this book is to document the rich history of Cape Canaveral through photographs of the facilities: of the rockets that flew there, the facilities that supported the launches and the people who worked there over the years.

Cape Canaveral is hardly abandoned or desolate today. The newest space boosters at the Cape are known as the Evolved Expendable Launch Vehicles (EELV), Boeing's Delta IV and Lockheed Martin's Atlas V rockets. These state-of-the-art vehicles are designed to perform most of the classified military and government (NASA) space missions that used to be performed at Cape Canaveral with the familiar Thor, Atlas and Titan launch vehicles.

Launch activity at Cape Canaveral is actually on the upswing according to Aviation Week magazine. Planners at the 45th Space Wing, the Air Force unit that operates the range at Cape Canaveral, report that at least 20 launches are booked in 2006, a marked improvement over the historically low total of 2005. There is a slight decrease forecast in 2007, but at least 24 launches are penciled in for 2008 and 2009, rising to 27 launches by 2010. This activity reflects a revival of commercial launch history in the world market, and the return of the Space Shuttle as a user of the range.

It is still too early to predict the future evolution of Cape Canaveral. Will it become the literal spaceport of science fiction stories, with regularly scheduled rocket service to orbiting cities, the moon and the planets? Or will it decline into irrelevance and eventually close its gates? For at least the decade to come, Cape Canaveral will continue to support the civilian and military space programs as it has for nearly 50 years. Long may it operate.

Joel Powell
February 2006

'Welcome to Cape Canaveral'

The sign at the main entrance of Cape Canaveral today.

The main gate of Cape Canaveral Air Force Station is located adjacent to Port Canaveral.

The logo for the 50th anniversary of Cape Canaveral was displayed at the main entrance to the base in 2000.

It was a different era: the original test range at Cape Canaveral in the 1950's launched multi-million dollar rockets and missiles with primitive vacuum tube technology and analog instrumentation - the digital revolution was at least thirty years in the future. The telephones were rotary (no such thing as a cell phone!) and there were no DVD machines, MP3 music players (just record players) or home computers. Computers were still extremely rare in 1950's America, and the ones that did exist filled entire rooms and had to be air-conditioned because the vacuum tubes they used generated so much waste heat. The maze of wiring in the picture at left is part of the telemetry receiving equipment at Cape Canaveral which had to be tended by crews of technicians to keep the receivers operating.

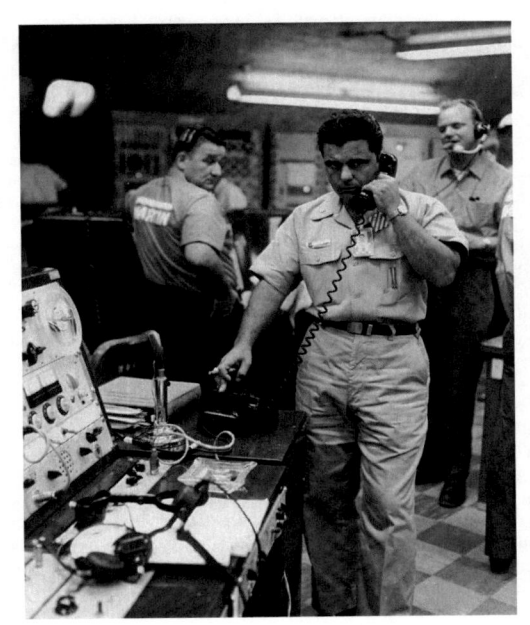

At left the Titan II project officer appears to be engaged in a heated discussion on the telephone, surrounded by the now antique knobs, switches and mechanical gauges of the pre-digital age.

CAPE CANAVERAL

On nearly deserted Playalinda Beach just north of Cape Canaveral, Florida, the soothing sound of the surf shushes constantly in the background as the pelicans glide in the thermals just off shore. The isolation is so complete that one can imagine Spanish caravels passing by on the horizon nearly five hundred years ago. Those courageous seamen could not possibly have imagined the remarkable vehicles that would rise into the heavens from a small spit of land on the La Florida coastline in the far-off future.

Spanish nobleman Juan Ponce de Leon, the discoverer of Florida, is credited by historians for being the first European to explore the Cape Canaveral area in 1513. The point of reeds (Canaveral) is what his cartographer called the spit of land between the Atlantic Ocean and what would later be known as the Banana River. The Canaveral name is said to be one of the two oldest geographical place names in the United States. Ponce de Leon did not get an enthusiastic reception from the natives on shore, and more than three centuries would pass until the first permanent American settlers arrived in the Cape Canaveral area sometime after 1840.

Cape Canaveral served as a vital dead reckoning landmark for European mariners returning home from the Caribbean. For several centuries after Ponce de Leon, the tricky currents off the Cape claimed many a sailing ship, and this prompted the United States government to build a lighthouse on the shoreline of Canaveral in 1847. After the Civil War the lighthouse was replaced in 1868 with a 164 foot wooden structure with an iron-plated exterior. The building was dismantled and moved to its present location in 1892-94 on a dune close to what is now one of Cape Canaveral's newest launch sites, Complex 46.

In the early days of Canaveral's missile launching career, the lighthouse was often used as an observation point for missile firings. The lighthouse was constantly mistaken by tourists for a rocket itself, which inspired the base's public affairs office to produce a humorous short film called "The Lighthouse That Never Fails." A hapless "airman" hangs on for dear life when the old lighthouse actually launches itself into space with a clever application of 1950's trick photography and missile launch footage. The lighthouse is still in service today for the Coast Guard.

When Nazi U-boats brought the Second World War to America's Atlantic shores in 1940 (prior to the Japanese attack on Pearl Harbor), the U.S. Navy established a seaplane patrol base 15 miles south of Cape Canaveral. The Banana River Naval Air Station was commissioned on October 1, 1940 and remained in service until April 1947. The runways and shuttered buildings of the Air Station would soon play host to the headquarters of the Joint Long Range Proving Ground in the opening days of the Cold War.

After World War II, captured German V-2 rockets were test fired by the Army at an inland test range at White Sands, New Mexico. Hemmed in by populated areas in New Mexico, Texas and Mexico, White Sands could only accommodate missiles with short ranges on the order of a few hundred miles. It was clear to Pentagon officials that a larger range was required to test fly intercontinental ballistic missiles that were being studied by the armed services.

The Joint Chiefs of Staff, the military command structure at the Pentagon, formed a committee to select a location for the new long range proving ground. A prime site in California was rejected because the President of Mexico refused to grant over-flight rights to the Americans, and the committee eventually settled upon an innocuous point of land on the eastern coast of Florida: Cape Canaveral in Brevard County. Aside from reasons of isolation and sparse population, the key reason to choose Cape Canaveral was Britain's willingness to allow missiles to over-fly Bermuda to reach planned downrange tracking stations.

The nearby islands of Grand Bahamas, Grand Turk, Antigua and far-away Ascension were ideal locations for downrange tracking stations, strung out in more or less a straight line downrange from Cape Canaveral all the way to the South Atlantic (Ascension). With the sites of the future tracking network secured, the Long Range Proving Ground was activated on October 1, 1949. The U.S. Air Force, just two years old, was responsible for operating the test site and downrange tracking stations when the facility was operational. Headquarters for the Proving Ground were established at the former Banana River Air Station, which was renamed Patrick Air Force Base on October 1, 1950. The name honored the first chief of the old Army Air Corps, General Mason M. Patrick.

Construction of the first crude launch facilities (and an access road into Cape Canaveral) commenced in May 1950. The first rockets scheduled to fly at the new Proving Ground arrived in June 1950 in a truck convoy from White Sands Missile Range: two-stage Bumper rockets that consisted of a modified V-2 first stage and a WAC Corporal upper stage. The two missiles were to propel the Corporal

(Above) The first blockhouse at Cape Canaveral (top) faces two of the original launch sites: the former Redstone site at Complex 4 at right with one of the split-roof shelters for the Bomarc interceptor, and Complex 3 at left where the first Bumper launch took place (the launch ring was located just below the Jason launch rail at the extreme left). The photo was taken in 1961.

(Left) Bumper rocket in position at the original launch site in July 1950. Note the former painter's scaffold at left that served as the service tower.

upper stages on essentially horizontal test flights, which could not be accommodated at White Sands or from the new Navy missile range at Point Mugu in California.

A deceptively simple concrete pad of 9,500 square feet was prepared for the Bumper firings. It concealed underground work passages leading to the pad and equipment rooms.

The site was later designated as Launch Complex 3 when two new launch pads were built in the same area in 1952. A temporary plywood launch control center (formerly a local bathhouse) was set up 500 feet from the launch site, which was later replaced by the first concrete "blockhouse" at the Cape (photo). Many years later, in the 1990's, Air Force missile museum volunteers uncovered the site of the control shack and launch stand for preservation as an historic site.

The first attempted launch at Canaveral on July 19, 1950 was also the first abort or scrub, the bane of missilemen and rocket scientists to this day. Bumper Number 7 fizzled on the launch pad, forcing the Air Force to haul it away to Patrick AFB for repairs. It was replaced by Bumper Number 8, which became the first rocket to launch from Cape Canaveral at 9:29 AM on July 24, 1950. The experiment failed when the Corporal stage did not ignite. Five days later Bumper 7 was completely successful, when the WAC-Corporal traveled 150 miles downrange and achieved the highest sustained speed in the atmosphere to date, in excess of 3,200 miles per hour.

After departure of the Army personnel back to White Sands, the Long Range Proving Ground was christened as a rocket launch facility, but Cape Canaveral was far from fully operational. Indeed several years of frantic construction activity were scheduled by the Air Force to build the infrastructure required to test long range ballistic missiles. Even as Air Force crews were trained in the art of missilry by firing Fairchild Lark missiles in the first year of Cape Canaveral's existence, construction crews were busy enlarging the runways at Patrick Air Force Base, installing underground communication cables and building roads throughout the 14,513 acre missile test facility.

Hangar Assignments at Cape Canaveral Industrial Area: Then and Now

Hangar	Contractor and Original Use	Current Use
A	at Patrick AFB	USAF
B	at Patrick AFB	USAF
C	original Cape hangar (cruise missiles)	display missile restoration
D	ABMA (Redstone, Jupiter)	USAF
E	North American (Navaho)	USAF (former IUS stage)
F	North American (Navaho boosters)	USAF
G	Northrop (Snark)	USAF
H	Northrop (Snark - later, upper stages)	USAF
I	Boeing (Bomarc - later, Minuteman stages)	USAF
J	Convair (Atlas)	Atlas to 2005
K	Convair (Atlas)	Atlas to 2005
L	Douglas (Thor - later, Delta)	NASA (life sciences)
M	Douglas (Thor - later, Delta)	Boeing/Delta II
N	re-entry vehicle processing	NASA
O	never built	
P	never built	
Q	never built	
R	ABMA/JPL (Jupiter, Juno I, Juno II, Mercury-Redstone)	USAF
S	Project Vanguard (later for Project Mercury)	USAF
T	Martin (Titan I, II)	USAF
U	Martin (Titan I, II)	USAF
V	never built	
W	never built	
X	never built	
Y	Lockheed (Polaris, Poseidon)	U.S. Navy
Z	never built	
AE	Chrysler/NASA (Saturn I, IB; also Ranger and Delta-class payloads)	NASA
AF	NASA (Saturn; early Gemini and Apollo spacecraft checkout)	NASA (shuttle SRB processing)
AM	NASA (OGO, Pioneer, ATS)	USAF
AO	JPL (Surveyor)	USAF

Minor facilities

AA, AB, AC	(various processing assignments)	USAF

(Left) The famous Cape Canaveral lighthouse, erected in 1868, is operated today by the U.S. Coast Guard.

(Below) The main entrance of the Cape Canaveral Headquarters Building at Patrick Air Force Base in the late 1950's.

Cape Canaveral was a miserable place to live and work in the early 1950's. Housing was so scarce in the tiny local towns of Cocoa, Cocoa Beach and Titusville (the entire population of Brevard County in 1950 was only 23,653 residents) that workers at the Cape had to live in tents until quarters were built at nearby Patrick Air Force Base. So rapid was the growth around the missile test center that the population of Brevard County grew to 111,435 people by 1960, an incredible increase of 371 percent. Cocoa Beach in particular became a classic boom town with a tight housing market and all the social problems that went with rapid growth (with a soaring divorce rate among other things).

According to the late Martin Caidin, an eminent aerospace writer who covered the early days of Cape Canaveral, the swamps and palmetto groves that dominated the terrain bred incredible hordes of hungry mosquitoes that made life all but unbearable for construction workers until the swamps were drained and mosquito spraying programs were instituted by the Air Force. There were also plenty of poisonous snakes, alligators and other critters (including the occasional Florida wildcat) to threaten unwary workers at "Capebase," as the facility was known to Air Force personnel.

All through 1951 the first major installation, the Central Control Building, was constructed at the north end of the base in what would become the Industrial Area. This was the nerve center for the range, coordinating all operational aspects of activity from communications to missile tracking and the all-important Range Safety function. Later the massive Technical Laboratory was built at Patrick Air Force Base. This was where all the masses of electronic data for each missile test was reduced and analyzed (recorded on endless streams of pink punched paper tape), and where all the miles of motion picture film and thousands of photographs of each test were developed and studied.

Construction also began in 1951 on the harbor now known as Port Canaveral (preparatory work was actually begun in July 1950). The channel and harbor were dredged for the Air Force near the south boundary of the base. The port was originally intended for logistical supply ships for the downrange tracking stations. Lack of rail lines in the vicinity of Cape Canaveral greatly impeded the delivery of vital supplies to the base. The new port became an essential collection point for sea-borne delivery of structural steel, concrete and other construction materials, with the added bonus of being literally right next door to the main gate of the base. The port was opened for business in 1953.

Additional wharf space was constructed for the U.S. Navy in 1958 to support Project Polaris. In 1960 the Fleet Ballistic Missile submarines began arriving at the harbor to inaugurate sea-launch trials of the Polaris missile. Today the submarine missile launches are the longest running activity at the base, now surpassing 45 years of continuous off-shore test firings.

Cape Canaveral in 1951-52 was still a Long Range Proving Ground without any long range missile systems to test. The first of several name changes was applied by the Air Force in June 1951: the Proving Ground became the Air Force Missile Test Center (AFMTC). The first real missile to appear at Canaveral was the jet-powered Martin Matador cruise missile in June 1951 that was test fired from truck-mounted launchers. The Matador had a range of only 620 miles. It was not until August 1953 that the first large ballistic missile, the 63 foot Redstone, was test fired by the Army off of a simple mounting ring on a concrete slab (pad 4), very close to the original Bumper launch site.

The larger Northrop Snark cruise missile (with true intercontinental range) began test firings in November 1952 at the Cape, and it was also launched from a mobile platform utilizing two large solid booster motor. The Air Force built a 10,000 foot asphalt runway in 1953 for Snark recoveries (and later, the X-10 pilot-less aircraft) called the 'Skid Strip,' named for the landing skids used in place of wheels on the Snark missiles. The first Snark recovery was successfully completed on 2 October 1956. The strip also serves as Cape Canaveral's airfield to the present day, where rockets and payloads have been delivered to Cape Canaveral since the mid 1950's. Pegasus boosters have also sortied from the strip.

A runway barrier system was installed at the strip in 1955 to prevent the North American X-10 (test beds for the Navaho cruise missile) from veering off the runway after touchdown. The Skid Strip was refurbished in 1987, and once again in 2004 to reduce the width to 200 feet with 25 foot shoulders, bringing the facility up to current Air Force standards.

By 1953 Pan American Airways was in charge of base security and the downrange stations. RCA held the contract for technical support on the base, as well as for the all-important photographic services.

In 1954 the Army built a new dual-use launch site for Redstone and its derivatives, Launch Complex 5/6, that shared a dedicated blockhouse. Complex 5/6 still utilized simple launch rings on a con-

(Top) The Technical Laboratory building for the range was located at Patrick Air Force Base. The photo shows the installation of a Thor missile for display outside the laboratory in 1961.

(Right) The Range Control Center located in the Industrial Area (seen here in 1966) was the nerve center of the Cape, monitoring all launches and connecting the tracking stations. The Range Safety Officer was located at this facility.

(Left) The range headquarters building at Patrick AFB as seen from Highway A1A.

crete apron, but the sites provided improved firefighting (water spray) equipment and permanent rails to move the servicing gantry to and from the launch pad. The difference between Redstone's pad 4 and the new facilities was the method of erecting the boosters. At pad 4 the Army used a truck-mounted hydraulic arm to position the rocket, while a large crane was used to place Redstone on the launch ring at pads 5 and 6.

The reason that the Army launch facilities were much simpler than the concrete hardstands of the Air Force Thor and Atlas missiles, according to General Bruce Medaris, commander of the Army Ballistic Missile Agency, was that Army missiles were static fired ahead of time in Huntsville, Alabama, while the Air Force conducted their static test firings right at the launch pad. An elevated launch stand could also provide a flow of cooling water to counteract the hot exhaust gasses generated by the larger Air Force missiles.

On February 1, 1956 the Redstone Arsenal field office responsible for Redstone flight testing was named the Missile Firing Laboratory, which was led by Dr. Kurt Debus from Huntsville, who later rose to direct the Kennedy Space Center in 1962. After hosting its first Redstone launch in April 1955, Complex 5/6 was utilized for 16 Redstone and 9 Jupiter IRBM test flights through to the summer of 1961. In 1960 pad 5 was converted for suborbital test flights of Project Mercury with modified Redstone vehicles, including the missions of astronauts Alan Shepard and Gus Grissom.

The North American Navaho would be the largest missile to be tested at the range to date, and the Air Force began construction of the elaborate elevated concrete hardstand for the rocket-boosted cruise missile in September 1953. Launch Complex 9 featured a unique fold-down erector gantry to service the missile. In 1955 a tactical-type launch facility (pad 10) was built adjacent to the Navaho hardstand. A small rail-mounted portable service tower was also installed at the second Navaho site. Although the mobile-launcher concept was exercised several times at pad 10 in May 1957 with a flight worthy vehicle, the facility was never used for a Navaho launch. All eleven Navaho test flights were conducted at pad 9, which was the Cape's first elevated ("monument" type) launch complex, a harbinger for the missiles to come.

The year 1956 was the start of an incredible period of expansion at the Cape Canaveral base, with the start of construction for no less than eight separate launch complexes intended for Thor, Atlas and Jupiter missiles and the Vanguard satellite launcher. Ground was broken for the first two Atlas complexes, pads 12 and 14, in January 1956, followed by start-up at Complex 17 in April 1956 for Thor. Work was underway at Complex 18 in June 1956 for the Vanguard project, along with a simplified tactical launch site for Thor, followed by the start of construction at pads 11 and 13 for Atlas in August1956. Work also started that year on the dual pad facility for the intermediate ranIge Jupiter (Complex 26), which was occupied by the Army in May 1957. The Army designated the site Vertical Launch Facility 26. The first Jupiter missile was launched from pad 26A on August 28, 1957. The concrete apron for the Bull Goose decoy missile (Complex 21/22) was also constructed near the original Bumper site close to the lighthouse.

The Cape's Industrial Area also saw a frenzy of construction in 1956 as the missile support hangars were erected in preparation for the new programs coming soon to the base for flight testing. Total cost for the hangars alone was $30 million in 1957 dollars. As many as twelve of the massive hangars, designated by letters of the alphabet, were completed in the area during 1956-57. The original hangar ('C') was built in the early 1950's near the lighthouse to support Matador testing, and was temporarily used by the Vanguard project until Hangar S was ready for them in early 1957. The Cape Canaveral workforce rapidly expanded to over 12,000 people by the late 1950's, a mixture of civilian contractors and military personnel.

Hangar D (for Redstone) and Hangars E and F for Navaho were ready before the completion of the Atlas, Thor and Titan facilities. All of these structures are still in use today, although only Hangar M for Delta II still supports an active launch vehicle.

Complex 14 for Atlas was completed first - in January 1957 - and supported the first Atlas launch on June 11, 1957, a fiery failure. It took three tries by manufacturer Convair to achieve the first successful Atlas A flight in December 1957. By April 1958 all four Atlas pads were activated, and would collectively host 83 research and development test flights through December of 1962. In May 1958 the name of the Cape Canaveral range was changed again to the Atlantic Missile Range (AMR).

Each individual Atlas complex had its own 60 foot diameter domed concrete blockhouse, whereas

The twin Thor-Delta pads are captured on film in 1961 (LC-17A center, LC-17B upper left) with the Army's Complex 26 at the top. Both gantries are in position at their respective launch pads to support a launch.

An aerial portrait of Launch Complex 20, recently activated for Titan I operations in December 1959. The large hydraulic lifting arms for the erector gantry are clearly visible. Compare the layout here with an aerial portrait of an Atlas facility (see p.30).

the Thor and Jupiter sites each consisted of twin launch pads that shared a single (rectangular-shaped) blockhouse. Launch Complex 17 has been in active service since December 1956 and is the longest serving launch site at the Cape. It currently supports the Boeing Delta II launch vehicle, one of the most reliable rockets in the world, which so far has failed only once since 1989. In 1997 the Air Force moved launch operations from the original Thor blockhouse at LC-17 to the 1st Space Launch Squadron Operations Building located about a mile from the Delta II launch pads. The move was planned before the low altitude destruction of Delta 241 in January 1997 (see Incidents and Accidents), which validated the decision to vacate the old blockhouse when poisonous fumes seeped into the interior after the explosion.

Complex 26 remained in service until January 1963, to support training launches of Jupiter by NATO allies Italy and Turkey, where the missiles were deployed overseas. The facility now serves as the site of the Air Force's Space and Missile Museum (along with Complex 5/6) adjacent to Complex 17. The museum was first opened to the public in 1966 in connection with Sunday drive-through tours of the base which had been initiated by the Air Force in December 1963.

Two new missile systems were due to make their debuts at Cape Canaveral by the end of the decade, Titan and Polaris, and so the year 1957 was dedicated to yet more massive construction projects. By the end of 1959 all eight of the Atlas and Titan pads lined up on the spine of Cape Canaveral to form the famous "ICBM Row." The Titan pads were distinguished by their individual fixed umbilical towers and tilting erector gantries that served to stack the missiles on the pad and provide access for pre-launch servicing.

The erector gantries were lowered hydraulically to ground level less than an hour before launch, compared to the rail-mounted Atlas gantries that were set back from the pad and moved by diesel engine to a fall-back position perpendicular to the pad. Work commenced on each of the Titan pads in 1957, and the first site (pad 15) was occupied by the Martin Company in April 1958. The first Titan I launch did not occur until February 3, 1959 at this site. LC-16 was activated in June 1958 and LC-19 was ready in July 1958. The final Titan pad was not developed as quickly as the other three. Labor troubles dogged the Titan construction project throughout 1958.

LC-20 was not occupied until July 1959, and did not support its first Titan I launch until July 1, 1960, a spectacular failure where the missile heeled over immediately after liftoff and was destroyed by Range Safety after only a few seconds of near-horizontal flight. Footage of the disaster appeared in the movie "The Right Stuff." Two of these pads, LC-15 and LC-16, were later utilized for the more powerful Titan II ICBM, which unlike the liquid oxygen /kerosene powered Titan I used storable hypergolic liquid propellants that ignited on contact.

The first Titan II flew from pad 15 on March 16, 1962. The Martin Company needed just 23 test flights to qualify Titan II to operational status, less than half the 47 flights needed to qualify Titan I. LC-20 was later modified to support the Titan IIIA space launch vehicle, which was actually the core stage of the Titan IIIC heavy lift booster that would be ready for flight test in June 1965.

Facilities for the diminutive 28 foot Navy Polaris Fleet Ballistic Missile (FBM) were developed rather more quickly on the south end of the base. Polaris was the first solid-fuel missile to be flight tested at Cape Canaveral. Work on Complex 25 was started and completed before the end of 1957, with an activation date of December 18, 1957. The first sub-scale development vehicle was launched from pad 25A on April 18, 1958. The first full-scale AX-1 Polaris prototype followed on September 24, 1958. Pad 25B was an elaborate reproduction of the submarine launch system for Polaris called the Ship Motion Simulator, built by Lowey-Hydropress.

The entire launch mount was moved hydraulically to represent the motion of the submarine or launch ship, and the missile was physically ejected from the launch tube by compressed air. A total of eight Polaris missiles were launched from the Ship Motion Simulator from August 1959 to August 1960. A further 25 Polaris missiles were launched from a surface ship, the USS Observation Island (EAG-154) that was also equipped with the complete compressed air launch system that ejected each Polaris from the submerged submarine's launch tubes (dubbed "Sherwood Forest" by submariners). A second Polaris launch complex (LC-29) was opened in July 1959 and remained operational until 1967, supporting 47 Polaris development launches. Two more pads were built at Complex 25 in 1967 for the new Poseidon missile, LC-25C and 25D. Trident I was later tested from a newly renovated pad 25C beginning in 1977.

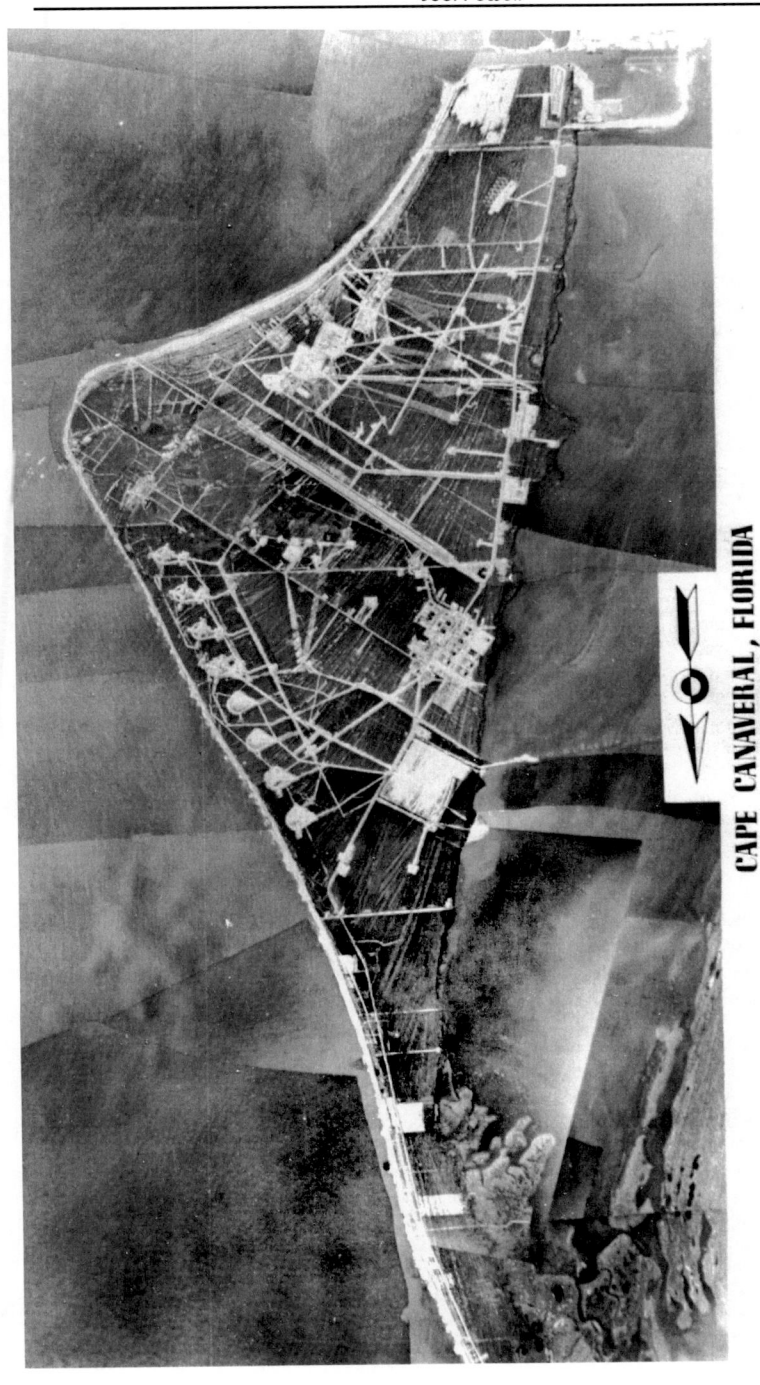

Cape Canaveral as it appeared from the air on February 15, 1957. The four Atlas pads are nearing completion on the central shoreline (on the right) while clearings for the four Titan pads to the left were just under way. Four clearings to the right of the Skid Strip runway are Complex 9, 17 and 18 and the Army's twin launch complexes 26 and 5/6 (left to right). The burgeoning Industrial Area is located to the left of the west end of the Skid Strip.

In 1959 the frantic construction pace continued as four launch sites were developed, including the relatively simple Pershing complex and blockhouse (LC-30) for the Army's solid fuel battlefield missile, and twin complexes for the Air Force's new Minuteman solid fuel ICBM. Groundbreaking also took place for a new launch site south of LC-14 for the new Atlas-Centaur space booster, as well as the first pad for the mighty Saturn rocket.

LC-30 was located in the vicinity of the Navy's Polaris complexes, while LC-31 and 32 for Minuteman were established on the former site of Navaho's pad 10. Each Minuteman site consisted of a buried silo and a separate above-ground pad for the initial test firings. Each complex had a distinctive beehive-shaped blockhouse covered in bags filled with concrete. The first Minuteman test flight occurred above ground from pad 31A on February 1, 1961. Pad 32A was a backup site and was never used for any launches.

Not only was the first flight successful, but it was also a demonstration of the new concept of "all-up" testing where all the upper stages of the missile were fired on each flight, rather than the incremental method of testing multi-stage rockets favored in the past. The first attempt to fire a Minuteman out of the underground silo at LC-32B resulted in a terrific explosion as the missile left the hole on August 30, 1961. Technical problems were overcome, and soon the missiles were exiting the silo smoothly on most flights (see *Incidents and Accidents* section for the exception to the rule). The two silos at Complex 31 and 32 were utilized for all three Minuteman variants through to the last Minuteman III test in December 1970, a total of 92 launches.

Work on Cape Canaveral's second dedicated space launch complex was underway in 1959 (LC-18A for Vanguard was the first) for NASA's Atlas-Centaur booster with a hydrogen-powered upper stage. Complex 36 was originally authorized for NASA's Atlas-Vega launcher with a liquid oxygen - RP-1 upper stage, but was transferred to the Centaur program when Vega was cancelled in December 1959. The 18 month construction job was completed in December 1960 and the first Atlas stage (104D) was erected for a seemingly endless series of fit and function tests on February 27, 1961. The first Atlas-Centaur came to grief during its maiden launch on May 8, 1962 and the vehicle was not declared operational by NASA until the AC-10 mission on May 30, 1966, the launch of Surveyor 1 to a successful soft landing on the Moon.

Construction of a second Atlas-Centaur pad, LC-36B, was started in February 1963 and was completed for NASA in July 1964. Distinguished by a much taller umbilical tower compared to the original, pad 36B was placed in caretaker status by NASA until pad 36A was wrecked in March 1965 by the explosion of AC-5 (see *Incidents and Accidents* section). Pad 36B was immediately activated, and supported the next Centaur flight test in August 1965. Complex 36 remained in service for nearly 40 years, with the last launch at pad 36A in August 2004 (the final conventional Atlas, a IIAS model). The final launch at pad 36B occurred in the fog on February 5, 2005 with a Russian-powered Atlas III vehicle.

One more landmark launch complex at Cape Canaveral also began life in 1959. Site clearance began in June 1959 for Complex 34, the impressive facility for Wernher von Braun's giant Saturn C-1 launch vehicle. NASA accepted the new Saturn facility during a brief dedication ceremony on June 5, 1961. The first Saturn I flight vehicle (SA-1) arrived by barge at Cape Canaveral on August 15, 1961. LC-34 also featured a 310 foot mobile service tower, the largest moveable structure in North America.

The first Saturn I was launched successfully on October 27, 1961 to test the S-1 first stage with its cluster of eight H-1 engines. In a curious engineering footnote, the first two Saturn launches were conducted without an umbilical tower. A tail service mast substituted for the umbilical tower until it was erected for the third launch at LC-34. A total of four Block I test flights (with dummy upper stages) were flown there before Saturn testing was transferred to an even larger launch complex built adjacent to pad 34: the twin launch pads of LC-37 to conduct the orbital flight tests.

Beginning in July 1964, Complex 34 was later converted to support Uprated Saturn I (Saturn IB) launches for the Apollo program beginning in July 1964. The open-air service structure was almost completely rebuilt, with enclosed hurricane gates, new service platforms and four environmentally-controlled work areas. The modifications were completed in August 1965. Pad 34 was where the tragic Apollo spacecraft fire occurred on January 27, 1967, killing the Apollo 1 crew of Grissom, White and Chaffee. Pad 34 was also the site of Apollo's triumphant return to flight in October 1968. Apollo 7 was the final Saturn vehicle to fly from Cape Canaveral - the remaining four Saturn IB's were flown from

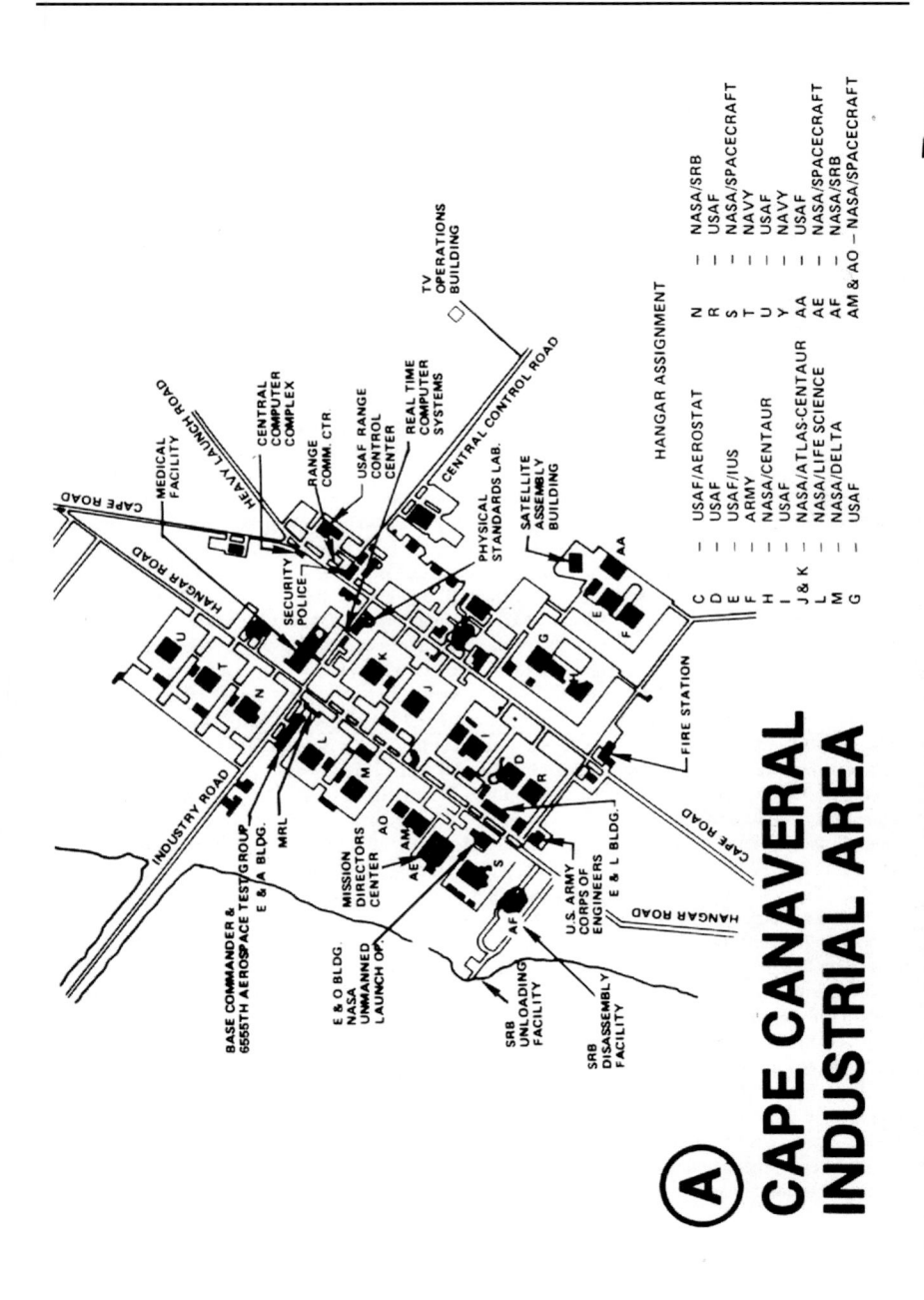

Map of Cape Canaveral Industrial Area

pad 39B at Kennedy Space Center, Canaveral's neighbor to the north, from atop an elaborate "milk stool" mounting on the former Saturn V pad.

The twin pads of Launch Complex 37 were accepted by NASA in August 1963. The first launch was performed with Saturn SA-5 on January 29, 1964 from pad 37B which successfully orbited the hydrogen-powered S-4 upper stage. All six Saturn I launches were made from the same pad; pad 37A served as a backup and was never used. The final three Saturn I's from LC-37 orbited the massive Pegasus meteoroid detection satellites, which deployed 49 foot "wings" mounting detector panels to assess the meteoroid hazard for Earth orbiting Apollo crews.

President John F. Kennedy made his third and final visit to Cape Canaveral on November 16, 1963. A week later the President was dead, victim of an assassin's bullets in Dallas. The new president, Lyndon Johnson, acted quickly to rename the missile center and its geographical location as Cape Kennedy (a decision that was reversed in 1973 by the Florida state government because the Kennedy Space Center name was deemed to be honor enough for the revered patron of NASA's Apollo moon program). In early 1964 the Atlantic Missile Range became the Air Force Eastern Test Range (AFETR), a name that was retained until 1979 when it was changed to the Eastern Space and Missile Center (ESMC) by the Air Force. In November 1991 the name was changed again to the 45th Space Wing.

"Rocket science" was new and exciting in the 1950's and 1960's for the residents of Brevard County and the workers at Cape Canaveral. Launch crews celebrated successful missile shoots with the famous "pool parties" at Cocoa Beach hotels. When somebody yelled, "Missile, Missile!" everyone would drop what they were doing and go outside to watch the launch, according to Cocoa Beach resident Nancy Yesecko in her award-winning documentary, "Growing Up With Rockets." The excitement lingered until the final Apollo moonshot in December 1972 from nearby KSC.

When federal funding began to decline at Cape Canaveral after 1969, the Air Force decided to downsize the test range as the advanced Minuteman III and Poseidon missiles neared the end of their test flight programs (with no major projects in the pipeline to replace these test programs on the range). The Air Force and their associated contractors had done a tremendous job at Cape Canaveral to establish the nation's deterrent forces against the Soviet Union, but after less than fifteen years since Thor and Atlas missiles first flew at the range, the job of developing ballistic missiles was largely complete.

Under Project DOWNSIZE, the Air Force transferred the big AN-TPQ-18 radars on Ascension and Grand Bahama Island to the Kwajalein test range in the Pacific, and the Mistram and Udop tracking systems on the range were deactivated. Grand Turk's command/destruct system was retired, and the tracking station on Eleuthera was transferred to the U.S. Navy in July 1971. Distant station 13 near Pretoria was shut down and returned to the government of South Africa in December 1969. The station on Grand Bahamas Island (GBI) was mothballed in June 1987 and turned back to the Bahamian government in January 1988. A new station was built at Jupiter Inlet, Florida in the 1980's to replace GBI: the Jonathan Dickenson Missile Tracking Annex, which also supports Space Shuttle launches from KSC.

The last missile test facility at the Cape was built beginning in February 1984 for the Navy's new 44 foot tall Trident II (D-5) submarine launched missile. The missile was taller and wider than the earlier Trident I, and was designed to be deployed (along with Trident I) from an entirely new class of missile firing submarines, the gigantic 560 foot Ohio-class boats with 24 launch tubes. Launch Complex 46 was built on the point area of the Cape near the famous lighthouse. The $30 million complex was designated the Trident II Development Flight Test Facility. The first land-based test of the Trident II was conducted on January 15, 1987.

After 19 test flights from LC-46, the Navy commenced submerged test launches from the USS Tennessee in March, 1989. On that first launch, water pressure damaged the first stage nozzle and the missile spiraled out of control (see Incidents and Accidents). LC-46 was converted to launch the solid-fuel Lockheed Athena space booster in the 1990's, but after only two launches the market for small launchers evaporated and the complex was placed on stand-by status in 1998, where it remains to the present day.

NASA also modified pad 37B to accommodate Saturn IB vehicles as a backup to LC-34, and two Saturn IB launches were performed from Complex 37 including Apollo 5 in January 1968, the first unmanned test of the Lunar Module in Earth orbit (using the AS-204 booster that would have performed the Apollo 1 mission had the tragic fire not occurred).

LC-37 was decommissioned in November 1971 and the service towers were demolished and sold

A fine study of Launch Complex 36, with the AC-4 vehicle undergoing preflight checkout on pad 36A in late November 1964. The recently completed facilities of pad 36B (foreground) were held in standby mode until needed, which turned out to be much sooner than anticipated (see the AC-5 explosion in the Incidents and Accidents section).

Facilities at Complex 34 for Saturn I dwarf the ICBM gantries stretching south to the horizon at Cape Canaveral in this high level view from the mid 1960's.

for scrap. The pad remained vacant until Complex 37 was chosen as the launch site of Boeing's new Delta IV Evolved Expendable Launch Vehicle in 1999. The site was redeveloped at a cost of more than $39 million over a period of two years. A new 100,000 square foot Horizontal Integration Facility (HIF) was erected on a 25 acre site adjacent to pad 37 to join together the modular Delta IV vehicles. Boeing chose a wheeled transporter to move the booster from the HIF and erect it on the pad. Up to three Delta IV Common Core booster segments could be stacked on the pad (side-by-side) to constitute the Delta IV Heavy version. A 330 foot service tower and a fixed umbilical tower were completed in 2000, and the site was dedicated during a ceremony on October 9, 2001. The first Delta IV thundered aloft on a commercial satellite delivery mission on November 20, 2002.

The final addition to the Cape Canaveral skyline in the mid 1960's was the Air Force's Integrate-Transfer-Launch (ITL) facility for the Titan IIIC heavy lift booster. Pads 40 and 41 were the largest and most elaborate launch facilities ever built at Cape Canaveral, and when the ITL was dedicated in early 1965 the base was essentially complete, taking the form and layout that would rapidly become famous around the world as America's space program began to reach for the Moon in the 1960's.

The Titan IIIC launch vehicle (originally developed for the cancelled X-20 Dyna-Soar program), was designed around a mobile launch process. The twin solid rocket motors were stacked on a rail-mounted transporter in the Solid Motor Assembly Building (SMAB) and were moved to the Vehicle Integration Building (VIB) for mating to the Titan III core vehicle. The entire stack was then moved to either pad 40 or 41 by twin Air Force locomotives. The difficult job of building the Titan III facility began in February 1963. Most of the ITL complex, located at the north end of the Canaveral base on the boundary with NASA's Kennedy Space Center, was built on 6.5 million cubic yards of hydraulic fill material dredged from the Banana River. Only the two launch pads (LC-40 and LC-41) were built on existing dry land.

After start of construction on August 12, 1963, foundations for each of the servicing buildings were built up with fill in the channel of the Banana River. The massive Mobile Service Tower and Umbilical Tower were completed at pad 40 in January 1965 and at pad 41 in April 1965. The first Titan IIIC vehicle with a dummy payload (including a copy of the base newspaper, the "Missileer") was launched on a totally successful orbital test flight from pad 40 on June 18, 1965.

The SMAB building was not yet operational, so the Air Force had to stack the boosters by crane right at the launch pad. There was no blockhouse at the Titan ITL installation. Instead the Air Force established a Control Room in the Vertical Integration Building (VIB), which was located a safe distance from the Titan launch pads.

The ITL complex went on to support 56 Titan launches through 1989, including eight upgraded Titan 34D versions (introduced in October 1982) and seven Titan IIIE-Centaur versions for NASA to launch the famous Voyager and Viking space probes. In 1985 the Air Force prudently decided to develop a Complimentary Expendable Launch Vehicle as an alternative to the Space Shuttle for defense missions, and the ITL facility was refurbished to accommodate the newly-named Titan IV booster. Complex 41, which was shut down after the last Titan-Centaur launch in 1977, was upgraded for Titan IV beginning in 1986. The maiden launch of the Titan IV was staged from pad 41 on June 14, 1989.

LC-40 was converted for Titan IV from 1990-1993. The Mobile Service Tower and Umbilical Tower were both demolished and completely rebuilt by Bechtel Corp. The most prominent feature of the rebuilt launch facilities was the four 350 foot lightning protection towers. In 1997 the final variant of the Titan family was introduced, the Titan IVB.

A new booster processing facility was required to handle the much larger solid rocket motor upgrade boosters (see Titan IV section), so the Air Force built the 320 foot tall Solid Motor Assembly Readiness Facility (SMARF) to replace the ITL's old SMAB building. SMARF, located just north of the existing SMAB booster facility, was used exclusively to process Titans after the last Titan IVA (K-17) flew in August 1998. The Titan IV's service record was marred by the failure of K-17 on August 12, 1998 and an earlier Centaur upper stage failure in 1993. K-17 exploded only 42 seconds after liftoff when the power supply to the rocket was momentarily interrupted by a short circuit, causing the vehicle to careen out of control. The final Titan at Cape Canaveral, B-26, departed pad 40 on April 29, 2005 to conclude nearly 40 years of service by Titan in the defense of the United States.

A diminishing mission model permitted the Air Force to shut down pad 41 in the late 1990's in order to build a new launch facility there for the Atlas V EELV that would be replacing the Titan IV.

Aerial view of the Titan IIIC Integrate, Transfer and Launch (ITL) site on the northern boundary of Cape Canaveral. The four-bay Vertical Integration Building is in the immediate foreground, and the Solid Motor Assembly Building is at center left. Lifting off from pad 40 is Titan IIIC-9, the MOL Heat Shield Qualification mission (test 0855) on November 3, 1966.

LC-40 under construction in March 1964. The umbilical tower is at left, while the Mobile Service Tower rises to the right. The slanted concrete structure is the throat of the flame deflector.

The Titan service and umbilical towers were both demolished by explosive charges on October 14, 1999. Lockheed Martin's new Atlas V installations were completed by the spring of 2001, which included a 292 foot Vertical Integration Facility (VIF) to stack the vehicles, located 1,800 feet from the rebuilt launch pad. Unlike Titan IV, the new Atlas V was designed to roll on rails to the launch pad from the VIF within 24 hours of the scheduled launch. This is called the "clean pad" concept. The first of many Atlas V rockets (AV-001) lifted off from pad 41 on August 21, 2002 with a commercial payload. Recent launches include the Mars Reconnaissance Orbiter in August 2005, NASA's first application of a major new launch vehicle since the Space Shuttle in 1981, and the Pluto New Horizons mission in January 2006.

Now that the ITL launch facility for Titan IV is dark and Launch Complex 36 for the original Atlas is shut down with the retirement of these venerable heritage boosters, Cape Canaveral will be quieter than it has been for many years, but the base will hardly be idle. Boeing's Delta II, arguably the world's finest launch vehicle, will continue to fly from LC-17, Cape Canaveral's longest serving launch site. Reports from Boeing indicate that the Delta II production line may be closed in a few years, but currently the manifest is booked through 2009. Delta IV and Atlas V will continue to fly commercial and military payloads - and the occasional mission for NASA - until at least the end of the decade, but budgetary pressure may force the Pentagon to choose one EELV or the other after 2010.

Cape Canaveral has played host to well over 3,200 launches since the first Bumper in July 1950. Most of the original launch pads are abandoned now (the famous "ruins" of Canaveral) and the job of testing ballistic missiles at the base has largely disappeared (except for the ubiquitous submarine launches!), but there remains the nearly 50 year legacy of space exploration launches to uphold. There are still many missions to send to Mars and the outer planets (to say nothing of national security payloads and commercial satcoms) from that innocuous spit of land on the east coast of Florida known as Cape Canaveral.

Joel W. Powell

LIST OF CAPE CANAVERAL LAUNCH FACILITIES
(including numbers not assigned or facilities not built)

LC assignments

LC	assignments
1	Mace, Snark
2	Mace, Snark
3	Bumper; Lark, Mace, Polaris FTV, Snark
4	Redstone, Bomarc
5	Redstone, Juno II
6	Redstone, Jupiter
7	NOT ASSIGNED
8	NOT ASSIGNED
9	Navaho
10	tactical Navaho, Draco
11	Atlas
12	Atlas; Atlas Agena
13	Atlas; Atlas-Agena
14	Atlas, Mercury-Atlas; Agena
15	Titan I, Titan II
16	Titan I, Titan II (later Pershing)
17	Thor, Delta (two pads)
18A	Vanguard, Blue Scout Jr.
18B	tactical Thor, Blue Scout I, II
19	Titan I, Gemini Titan
20	Titan I, Titan IIIA
21	Mace hard shelter
22	Mace soft site
23	proposed site of ship-launched Jupiter (never built)
24	proposed site of ship-launched Jupiter (never built)
25	Polaris, Poseidon, Trident I (four pads)
26	Jupiter, Juno I (two pads)
27	proposed for Navaho II (never built)
28	NOT ASSIGNED
29	Polaris
30	Pershing I
31	Minuteman I, II, III
32	Minuteman I, II, III
33	NOT ASSIGNED
34	Saturn I, later Saturn IB
35	NOT ASSIGNED
36	Atlas Centaur (two pads)
37	Saturn I, IB (two pads); later Delta IV
38	proposed dual-use Atlas-Centaur/Agena (never built)
39	Saturn V, Shuttle (at KSC)
40	Titan IIIC, Titan IV
41	Titan IIIC, Titan IV; later Atlas V
42	proposed Titan IIIC (never built)
43	weather rockets until 1984
44	NOT ASSIGNED
45	Roland missile (cancelled)
46	Trident II; later Athena
47	weather rockets since 1984

CAPE CANAVERAL LAUNCH TOTALS BY YEAR
1950 to 2005

Year	Number
1950	5
1951	22
1952	41
1953	59
1954	75
1955	72
1956	94
1957	112
1958	147
1959	168
1960	206
1961	186
1962	151
1963	150
1964	105
1965	73
1966	77
1967	50
1968	47
1969	68
1970	66
1971	55
1972	34
1973	35
1974	40
1975	58
1976	44
1977	54
1978	62
1979	51
1980	63
1981	38
1982	53
1983	55
1984	24
1985	26
1986	28
1987	47
1988	34
1989	23
1990	34
1991	23
1992	42
1993	19
1994	26
1995	32
1996	29
1997	29
1998	26
1999	23
2000	22
2001	20
2002	18
2003	19
2004	14
2005	9

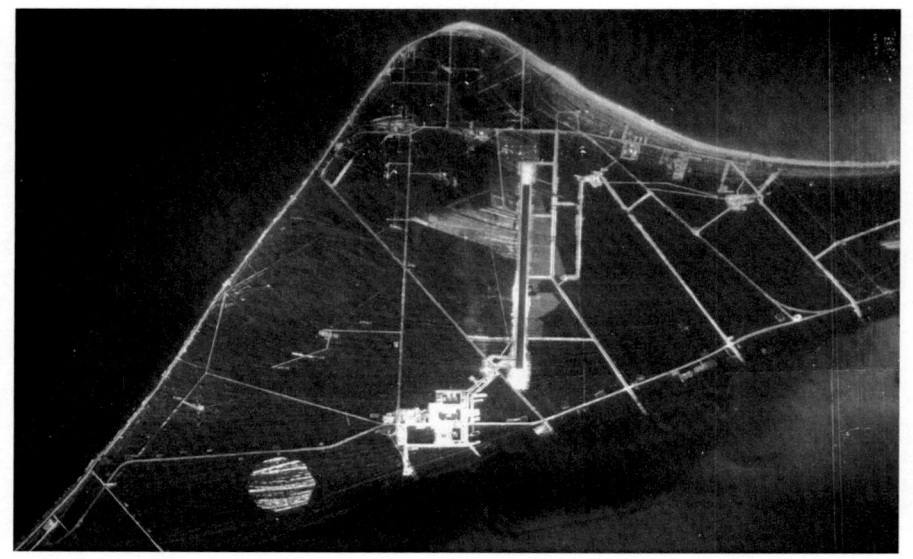

Two aerial survey photos from the early days of Cape Canaveral. below is the Cape as it appeared on March 17, 1951 where the main access road is the most prominent feature. The original Bumper launch pad is within the rectangular area marked near the tip of the cape. Above is the base as it appeared on December 2, 1954. Below the Skid Strip stands out in the center of the peninsula. Complex 9 for Navaho is the small clearing to the right of the eastern end of the Skid Strip, connected by two roads to the new Redstone site, Complex 5/6.

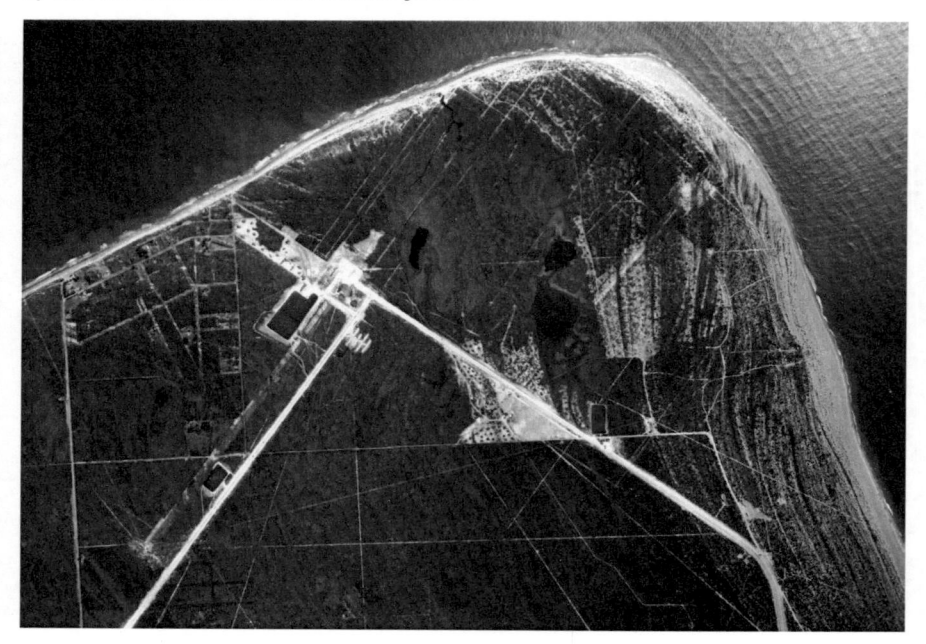

An aerial portrait of the first operational launch sites [not 'missile' sites] in 1957 at Cape Canaveral. In the foreground is Navaho's pad 10 (left) and the pad 9 hardstand; above it is Launch Complex 18 for Vanguard and the tactical Thors; just above that is Complex 17 for Thor, and the clearing (top) is the Army's Vertical Launch Facilities for Jupiter and Redstone.

(Right) Detailed aerial view of Complex 26 (twin pads in the lower portion) and Complex 5 and 6 in the upper portion (pad 5 is at the far end) in a photo dated May 14, 1958. Each Complex has its own gantry.

(Below) The Army ABMA sign at "Vertical Launch Facility 26" in May 1959 prior to the flight of space monkeys Able and Baker.

Atlas Launch Sites Under Construction

(Right) One of the Atlas launch stands takes shape at Cape Canaveral in 1957. The launch mount will be installed on the left-hand side.

(Center) An aerial view of Atlas Launch Complex 11 as it nears completion.

The circular blockhouse (below) is complete externally, and is located 800 feet from the launch stand, which is visible on the right.

(Top) The view north at Cape Canaveral from the top of the gantry at Launch Complex 11 (Atlas) across to pad 12 and 13 (with the construction crane at the new gantry tower, circa 1963).

(Below) The long ramp that leads up to the launch mount is shown here for the Atlas pad at Complex 11. The other three Atlas pads line up on the horizon (LC-12, 13, 14).

(Above) Atlas 13A accelerates rapidly in the early moments of this successful test flight on February 7, 1958 from pad 14 at Cape Canaveral.

(Right) The Cape Canaveral fire department has been an integral part of missile testing since the early days of the Long Range Proving Ground. Fire fighting apparatus attended every missile launch at the base, and the firemen were responsible not only for the safety of the base personnel but also were charged with helping to clean up after missile firing accidents and to prevent brush fires from spreading in the wake of a missile crash.

(Left) A familiar sight at Cape Canaveral for many years was the Pan American Airways (PAA) security guards manning the checkpoints at every launch complex, controlling access to the base at the main gate and patrolling the vast territory of the missile test center. In the photo a PAA guard converses with a member of the Atlas workforce at pad 14 about the Atlas-Agena rocket on the pad, soon to place a Gemini target vehicle into orbit in 1966.

(Right) Smoke clears from a static firing test for the Gemini Titan 1 launch vehicle at Launch Complex 19, as seen from the air in January 1964.

(Left) A Titan I ICBM lifts off from pad 19 at the Cape in 1962, as seen from the 310 foot Saturn gantry at pad 34. Pad 20 is located in the foreground, and five Atlas pads line up in a row to the horizon (at right).

(Right) The Navy's Polaris area at the south end of Cape Canaveral. Complex 25 dominates the center of the picture, with pad A on the right, and pad B, the Ship Motion Simulator, on the left, adjacent to Complex 29 (see below). At the top of the picture is the Navy Industrial Area with missile hangar Y.

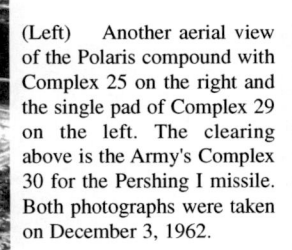

(Left) Another aerial view of the Polaris compound with Complex 25 on the right and the single pad of Complex 29 on the left. The clearing above is the Army's Complex 30 for the Pershing I missile. Both photographs were taken on December 3, 1962.

(Above & Right) Before-and-after shots of the original Navaho area (left) and the Minuteman area (right) that was superimposed over the existing Navaho pad 10 facility. Each Minuteman site consisted of one flat pad and one underground launch silo.

(Right) Low level view of the Minuteman area on April 7, 1961 looking north toward the distant Cape Canaveral Industrial Area. Complex 31's silo dominates the left side of the picture and the Complex 32 flat pad is situated on the right. Many years later the remains of the shuttle Challenger would be buried in both Minuteman silos (see Monuments section).

(Left) An unusual aerial view of Cape Canaveral (looking south down "ICBM Row") captures a Saturn I rocket in flight shortly after liftoff from pad 34, likely SA-3 on November 16, 1962.

(Right) The massive facilities for the orbital flight tests of Saturn I at Complex 37 near completion at the Cape in the latter stages of 1962. The other Saturn facility, Complex 34, lies in the immediate background.

(Left) The new Delta IV launch site at Complex 37 (pad B, at right) is located at the former Saturn I facility in the northern portion of Cape Canaveral. The remains of Saturn pad 37A (which was never used) still stands to the left of the Delta IV service structure.

(Right) Atlas V now occupies the site of the former Titan IV facilities at pad 41, located on the northernmost boundary of Cape Canaveral. Atlas V is prepared for flight in the Vertical Integration Facility (VIF) at the far left and is transported to the pad less than 24 hours before liftoff.

The 45th Space Wing

The 45th Space Wing is the present day host organization for the Air Force Space Command's Eastern Range at Cape Canaveral Air Force Station. The wing manages the far-flung assets of the Range from Cape Canaveral to Ascension Island in the South Atlantic, extending a full 10,000 miles to the Indian Ocean. The Range also includes Antigua Air Station, one of the only original stations remaining from the Cape Canaveral missile test range established in the 1950's. The other two installations for space launch tracking and support are the Jonathan Dickenson Tracking Annex at Ponce de Leon, and Malabar Tracking Annex south of Patrick Air Force Base.

In one of the periodic reorganizations that characterize American military operations, the 45th Space Wing was established on November 21, 1991 to replace the Eastern Space and Missile Center that had been in charge of the range since 1979. The 45th Space Wing traces its origin to the Joint Long Range Proving Ground that was set up at Cape Canaveral in 1950. Within a year the name was changed to the Air Force Missile Test Center, which was renamed the Air Force Eastern Test Range in 1964.

Three separate units within the Space Wing handle the day-to-day operations of the range: the 45th Operations Group, the 45th Launch Group and the 45th Mission Support Group which maintains facilities at the Cape and Patrick Air Force Base. Most of these responsibilities were handled by the legendary 6555th Aerospace Test Group at Cape Canaveral prior to the inception of the 45th Space Wing. This unit was formed in May 1951 as the 6555th Guided Missile Squadron (later Missile Wing) to test the Matador missile at the Cape, and was later redesignated the 6555th Test Wing in December 1959 when the unit was responsible for routine test operations at the base. The 45th Space Wing is commanded by a Brigadier General (see list of Commanders).

The 45th Space Wing sign at the main entrance to Cape Canaveral Air Force Station.

List of Commanders

Joint Long Range Proving Ground, LRPG Division, AFMTC, AFETR, ESMC & 45th Space Wing Commanders

Colonel Harold R. Turner	1 Oct 49 - 9 Apr 50
Major General William L. Richardson	10 Apr 50 -31 Jul 54
Major General Donald N. Yates	1 Aug 54 - 4 May 60
Brig General William L. Rogers	5 May 60 - 30 May 60
Major General Leighton I. Davis	31 May 60 - 1 Jan 64
Brig General Harry J. Sands, Jr.	2 Jan 64 - 16 Jul 64
Colonel Elmer W. Richardson	17 Jul 64 - 11 Aug 64
Major General Vincent G. Huston	12 Aug 64 - 4 May 67
Major General David M. Jones	5 May 67 - 31 May 73
Major General Kenneth R. Chapman	1 Jun 73 - 12 Aug 74
Colonel Dan D. Oxley	13 Aug 74 - 24 Aug 74
Brig General James H. Ahmann	25 Aug 74 - 24 Feb 75
Colonel Dan D. Oxley	25 Feb 75 - 5 Apr 75
Brig General Don M. Hartung	6 Apr 75 - 31 Jan 77

** Detachment 1 SAMTEC Era Commander: Colonel Oscar W. Payne **

Colonel John S. Burklund	1 Oct 79 - 30 Apr 81
Colonel Marvin L. Jones	1 May 81 - 13 Dec 84
Colonel Nathan J. Lindsay	14 Dec 84 - 24 Jun 86
Colonel John W. Mansur	25 Jun 86 - 12 Aug 87
Colonel Lawrence L. Gooch	13 Aug 87 - 27 Mar 89
Colonel Roy D. Bridges	28 Mar 89 - 26 Jan 90
Colonel John R. Wormington	27 Jan 90 - 22 Sep 91
Brig General Jimmey R. Morrell	23 Sep 91 - 29 Jun 93
Brig General Robert S. Dickman	30 Jun 93 - 23 Jan 95
Brig General Donald G. Cook	24 Jan 95 - 27 Aug 95
Brig General Robert C. Hinson	28 Aug 95 - 26 Mar 97
Brig. General F. Randall Starbuck	27 Mar 97 - 19 Aug 99
Brig. General Donald P. Pettit	20 Aug 99 - 6 Jun 02
Brig. General J. Gregory Pavlovich	7 Jun 02 - 25 Aug 04
Brig. General (Sel) Mark H. Owen	26 Aug 04 -

38 GO FOR LAUNCH

Part I

MISSILE TEST CENTER OPERATIONS

BUMPER

The first two rockets to be fired at Cape Canaveral in July 1950 were modified German V-2 missiles with WAC-Corporal upper stages (built by the Jet Propulsion Laboratory in Pasadena, CA), bearing the curious name of Bumper. Bumper 7 and 8 were tasked to fly essentially horizontal speed runs in the atmosphere with the Corporal upper stages, a mission chosen to inaugurate the new Joint Long Range Proving Ground because it could not be accommodated at the inland test range at White Sands, New Mexico. The first rocket launch at Cape Canaveral was Bumper 8 on July 24, 1950 (the Corporal stage failed to ignite), followed by Bumper 7 on July 29, 1950 which reached 3,270 mph, a record velocity at the time.

The first rocket to launch from Cape Canaveral, Bumper 8, rises from a clearing at the new Long Range Proving Ground under the watchful eye of news photographers on July 24, 1950.

LARK

Lark was the second missile to be flown at Cape Canaveral beginning in October 1950. Originally designed as an anti-aircraft missile for the U.S. Navy at the end of World War II, Lark was employed by the Air Force as a training vehicle to provide missile firing experience for Bomarc crews that would later operate at Cape Canaveral. The missile was 173 inches long and 18 inches in diameter. There were 40 Lark launches conducted at the Cape from 1950-53.

The second Lark missile (GM-226) is readied for firing on its zero-length launcher at the Long Range Proving Ground at pad 3 on October 26, 1950.

MATADOR AND MACE CRUISE MISSILES

Work on the first major U.S. cruise missile program started in 1946 by the Air Force and Martin Aircraft as a mobile weapon under the designation XB-61. Prototype flights started in 1949 at White Sands in New Mexico and were moved to Cape Canaveral in 1951 where 18 red-painted Matadors were flown to initiate the test program. Flight testing lasted until 1961, but the Matador was declared operational in Germany in 1954 as improvements (folding wings) and a new guidance system were tested. By 1962 the last of the Matadors in Asia and Europe were replaced by the improved Mace vehicle, which had commenced flight testing at the Cape in 1960. The Mace was not mobile, utilizing hardened shelters (see photos) to enhance its capability. The ultimate Mace B version was 44 feet long with an 18 foot wingspan, carrying either a conventional or nuclear warhead to a range of 1,500 miles.

Matador on a mobile trailer/launcher in March 1954 at Cape Canaveral.

A Matador under full thrust of its Allison J-33 jet engine is propelled off its mobile launcher in March 1958 by a single solid fueled booster before heading downrange.

The advanced Mace B is launched from a hardened shelter at Launch complex 21 similar to the operational sites it would be deployed at in Germany and Okinawa. The date was Sept. 21, 1960.

A Mace missile undergoes preflight checkout inside one of the slant silo facilities at Launch Complex 21 at Cape Canaveral in early 1964.

A proud group of airmen pose with their Mace missile inside one of the slant silo facilities at Complex 21 prior to launching their missile in 1963.

(Right & Top opposite) Two views of the concrete shell of the old slant silos for the Mace missile, LC-21, built on the concrete apron of the Old Bull Goose site in 1959-60, located near the lighthouse at the tip of the Cape. Mace missiles were test fired from the concrete shelter representing fixed launch installations at overseas bases in the 1960's.

SNARK

The Northrop Snark was the first intercontinental missile (ICM) developed by the U.S. Air Force, propelled by a powerful jet engine A, prototype version called the N-25 was used in the initial flight tests at Cape Canaveral that started in 1952. The standard test vehicle was the N-69 version with an Allison J-71 engine, later to be replaced by the advanced 10,500 pound thrust J-57 powerplant. Two Allegany Ballistics Laboratory solid fuel motors generating 105,000 pounds of thrust boosted the N-69 Snark vehicles into the air. Snark had an erratic test record, with one vehicle that disappeared into the jungles of Brazil in December 1956 and was not found until 1983. Some Snarks actually made round trips downrange and were recovered at the Cape's Skid Strip. Snark was 67 feet in length with a wingspan of 42 feet and a body diameter of 5 feet. The Snark was designed to carry the W-39 nuclear warhead. The Strategic Air Command maintained 30 operational Snarks at Presque Isle, Maine beginning in 1960 but in March 1961 Snark was declared obsolete by the Pentagon and was withdrawn from its brief service.

A dramatic view of a red and white striped Snark ICM being launched from the mobile transport truck at Cape Canaveral.

A Snark prototype is readied to be trucked to the launch site from its USAF hangar. This vehicle has the range-extending fuel drop tanks mounted under each wing.

With the Pratt and Whitney J-57 jet engine at full power, a Snark at Pad 2 is ready for the two 130,000 pound thrust solid motors to boost it into the air where flight can be sustained.

Engine run up prior to Snark launch from Pad 2. Note the looming gantries to the northeast for the successor to Snark: The Atlas Intercontinental *BALLISTIC* missile.

With a sudden impetus provided by the two solid booster motors this Snark was quickly airborne on March 3, 1960. The Snark hangar is visible in the background.

BOMARC
Bomarc was developed to counter the Soviet Union's potentially nuclear-armed heavy bomber force in the mid-1950's by destroying the aircraft at long ranges using a conventional or nuclear warhead. The Boeing Company was selected as prime contractor and started work in 1949 together with the Michigan Aerospace Research Center (hence the name Bomarc). The early Bomarc A version was propelled to ramjet ignition altitude by a liquid fuelled 35,000 pound thrust rocket engine. The ramjets were ignited at high altitudes and speeds that resulted in a high rate of approach to the enemy aircraft. Due to the complexities of the liquid propellant rocket engine, a 50,000 pound thrust solid motor was introduced to boost the newer Bomarc B missile, which had an 18 foot wingspan and weighed 8 tons at liftoff. Range was just over 400 miles. Early tests started in 1952 and the first complete prototype flew at Cape Canaveral in 1955. In 1957 the Air Force signed the first contracts with Boeing for the A model, which became operational in 1959. The follow-on B model was started in 1958 and became operational in mid-1961. A total of 140 Bomarcs were deployed on the U.S. eastern seaboard while 242 B-models were located farther north (including sites in Canada). After 1964 Bomarc numbers declined and by 1972 the missile was out of service. It was the only surface to air anti-aircraft missile (SAM) ever developed by the U.S. Air Force.

Bomarc dual-launch operation at LC-4 on January 27, 1959 - note the 'split-roof' shelter that enclosed the missile lifting off at left. (Below) Spectacular close up of Bomarc B liftoff at LC-4.

Bomarc A ignition at LC-4 circa 1957.

BULL GOOSE AT LC-21, 22
Bull Goose (XSM-73) was the developmental version of the Air Force's abortive Goose electronic countermeasure and decoy drone aircraft that was tested at Cape Canaveral in 1957-58. Twenty launches of the jet-powered test vehicle, built by Fairchild, were conducted from concrete aprons at Launch Complex 21 and 22. The 38 foot delta planform drone was blasted into the air by a small solid propellant booster. The flight test program was terminated in Dec. 1958 and the overall program was cancelled in 1959 without ever having been deployed operationally.

(Right) The third live Bull Goose test launch is conducted from Complex 22 at Cape Canaveral on September 26, 1957.

(Right) Artist's concept of Bull Goose decoy drone in flight.

NAVAHO

The Navaho rocket-boosted cruise missile was developed by the U.S. Air Force to deliver nuclear warheads over intercontinental distances before the advent of the Atlas ICBM in the late 1950's. The unsuccessful Navaho missile was cancelled in July 1957 following a ten year development effort, after contributing vital rocket engine and guidance technology to the next generation of American missiles. Built by North American Aviation, Navaho was 81 feet long and the twin-barreled rocket booster developed 325,000 pounds of thrust, the most powerful American rocket vehicle of its day when it debuted in 1956. The vehicle flew eleven times from 1956 to 1958 from pad 9 at the Cape, but the farthest the ramjet-powered cruise missile flew was 2000 miles downrange before flaming out. Navaho's unflattering nickname at the Cape was Never Go, Navaho.

(Right) Classic Navaho liftoff portrait of the final launch in November 1958. The vehicle broke up at about 60 seconds after liftoff, an ignominious end to the Research In Supersonic Environment (RISE) project, Navaho's last flight test assignment.

(Below) Only the nose and wingtip of the Navaho cruise missile is visible inside the unique erector gantry that positions the vehicle pivots the vehicle into position atop pad 9 for firing. In the background is the Vanguard launch site (pad 18A).

(Below Right) Navaho rocket booster minus the cruise missile is erected at pad 9 early in 1957 to conduct a static test firing of the rocket engine, a standard procedure in the early days of rocket development.

(Right) A portable tactical launcher for Navaho was installed at nearby pad 10 for shakedown tests in May 1957, but the vehicle was never actually launched from this facility. A mobile crane was used to assemble the vehicle on the portable launcher in the open air.

(Left) The massive ramjet-powered Navaho cruise missile is raised to launching position within the erector gantry mechanism at pad 9. The rocket booster is hidden behind the winged missile.

(Right) Navaho missile stands ready at pad 9 just prior to engine ignition. There was no umbilical mast at pad 9. The lattice structure in the background is a weather shelter for the Navaho workers.

(Left) The moment of liftoff for the first Navaho missile in November 1956. The flight lasted all of 26 seconds before the vehicle broke apart due to a guidance system malfunction. Before the advent of Atlas in 1957, Navaho was the largest vehicle to be flown at Cape Canaveral.

(Right) Piggybacking on its powerful rocket booster, the next-to-last Navaho vehicle soars off the launch stand in September 1958. The vehicle broke up shortly after booster separation, a particularly vulnerable portion of the Navaho flight path, learned the hard way during previous launches.

(Left) Navaho booster 009 is erected into firing position on the portable launcher at pad 10 (note the maw of the flame deflector at bottom right). Navaho 009 was later launched from the pad 9 concrete hard stand in August 1957.

REDSTONE AT PAD 4

Redstone was the Cape's first "large" rocket tenant (not that 69 feet is large by today's standards), ironically an Army missile test flown at an Air Force installation. Redstone was essentially an "Americanized" version of the German V-2 from World War II, developed by Wernher von Braun's rocket team at the Redstone Arsenal in Huntsville, Alabama. The Army established their Redstone launch site at Cape Canaveral in 1953 where the first test flight occurred on August 20. The first six Redstone launches departed from the original pad 4 facility (which was then assigned to the Bomarc program by the Air Force) from August 1953 to February 1955 before the Army transferred to their new Redstone launch complex (pad 6) in April 1955, which they dubbed the Missile Firing Laboratory. A total of 25 Redstone research and development and tactical prototypes were launched from Cape Canaveral from 1953 to June 1961, not counting the so-called Jupiter A versions launched to test components of the Jupiter intermediate missile (see P.52).

(Right) Heavily retouched Army photograph of what is reputedly the first Redstone launch in August 1953 from pad 4.

(Below) Unidentified Redstone vehicle just after erection at pad 4 in preparation for one of the early test flights circa 1954.

(Below) Redstone vehicle RS-7 (code RSXI) was the third launch from the new Vertical Launch Facility (pad 6) on August 30, 1955.

Redstone pioneered pathway to space

August 20 marks the 40th anniversary of the first launch of the Redstone missile. The 69-foot-tall vehicle traveled a mere 8,000 yards after liftoff from Pad 4 on Cape Canaveral Air Force Station, but the course it steered into history extends much further.

Originally designed as America's first ballistic missile, the Redstone's greatest contributions were peaceful in nature. The first U.S. satellite to orbit the Earth, Explorer I, was launched atop a modified Redstone commonly called Jupiter C in 1958. Another historic first occurred three years later, when a modified Redstone carried the first American into space.

The Redstone legacy extended over several generations of engine development to the most powerful rocket ever flown, the Saturn V which hurled Apollo spacecraft to the moon.

Redstone-descended engines powered the first of the Saturn V's three stages. Some 138 feet tall and 33 feet in diameter, the first stage dwarfed its remote ancestor. Its five powerful engines were built by the same contractor, Rocketdyne, and gulped the propellant combination – kerosene and liquid oxygen – used for the Thor and Atlas engines further up in the evolutionary chain. Redstone burned alcohol and liquid oxygen.

Saturn V liftoff thrust of 7.5 million pounds exceeded the 78,000 pounds of Redstone by more than 96 times.

"It is doubtful if any other single development program has ever produced such a rich harvest," observed Dr. Kurt Debus, who led the Redstone launch team for the U.S. Army. This group of 37 technicians and engineers eventually became the nucleus around which Kennedy Space Center was formed, and Debus became its first director.

The Redstone program was conducted at the cape from 1952 through 1960.

READIED FOR FIRING at Launch Pad 4 on the Cape is Redstone 4. The vertical service tower was the first of its kind and was actually a specially modified oil derrick mounted on rails. Liftoff occurred on Aug. 18, 1954.

A REDSTONE is within seconds of lif During the pre-stage sequence, a lau team member called the "flame obser watched the color of the alcohol/li oxygen exhaust flame to make cer that it had the correct mix of fuel oxidizer. "Main stage" firing followe positive determination of the propel mix.

BROUGHT TO THE PAD horizontally positioned on a combination transporter/erector, the Redstone was then raised to a vert position on a firing stand.

The Kennedy Space Center newsletter featured the Redstone missile in 1993 on the occasion of the 40th anniversary of the first launch at Cape Canaveral.

REDSTONE AND JUPITER A

The pace of Redstone flight testing was accelerated in 1955 when the Army Ballistic Missile Agency (ABMA) activated the new Vertical Launch Facility (the twin pads of Complex 5/6) at Cape Canaveral. A four year research and development test schedule was drawn up by the Army to qualify the Redstone for deployment in Europe with nuclear warheads. The series of 14 developmental firings was concluded on November 6, 1958 with a successful flight. In March 1956 a slightly different Redstone configuration made its first appearance at Complex 5/6: the Jupiter A, intended to test components of the Jupiter intermediate range missile. To reserve launch time slots on the Cape Canaveral range for the tests, the Army resorted to a bit of subterfuge (using the Jupiter A name) to convince the brass that this program was related to the higher priority Jupiter IRBM. A total of 24 Jupiter A missiles were fired by the Army at pads 5 and 6 through June 12, 1958. Finally there were 10 test firings at the Cape of the Block I and II tactical versions that represented the operational missile deployed in Europe. The final Redstone at Cape Canaveral was launched on June 27, 1961.

(Below) Redstone number 57 (code SI) is prepared for launch at pad 6 on November 5, 1958.

(Above) Dramatic close up of Redstone number 46 (code SV) just after lift off February 11, 1958 on a routine test flight from Pad 6.

(Above) Night launch of tactical Redstone number 2004, one of only two Redstones to be launched from pad 26A, on August 4, 1959.

(Right) Tower rollback for Redstone number 46 (SV).

PERSHING 1
Pershing I was a two-stage solid fuel battlefield missile developed by the Army Ballistic Missile Agency, designed to replace the Redstone for overseas deployment in Europe in the 1960's. The 35 foot Pershing I was built by Martin Orlando and was flight tested at Complex 30 at the Cape. The first of 61 test launches was conducted on February 25, 1960. Pershing I had a range of 460 miles. Later, Pershing IA Follow On Tests (FOT) by Army troops were conducted under tactical conditions at LC-16 from May 1974 to October 1983 for a total of 88 firings.

Pershing I R&D missile number 316 is prepared for the first launch from a Transporter - Erector Launcher (TEL) at Complex 30 on August 22, 1961 (test 3500).

(Right) The first Pershing missile is launched from Complex 30A on February 25, 1960. It was successful.

(Left) A Pershing IA Follow-On Test (FOT) was conducted at pad 16 on October 13, 1983, one of four Pershings launched that day from the Cape.

PERSHING II

Pershing II was an evolutionary improvement of the Pershing I missile, with a range of 1,100 miles, featuring improved propulsion elements and terminally guided re-entry vehicles. The Pershing II flight test program got under way at Launch Complex 16 at Cape Canaveral on July 22, 1982 (a failure). Martin Marietta conducted the first two launches, and two Army Field Artillery units conducted the remaining test and training launches through May 1988, for a total of 49 launches at LC-16. All Pershing IA and Pershing II missiles were eliminated by the INF Treaty between the United States and the Soviet Union on June 1, 1988, and most of the missiles were destroyed by the Army (crushing and burning). Pershing II at 34 feet 9.7 inches was two inches taller than Pershing I with a larger diameter.

The twelfth Pershing II missile is launched from a portable launcher at Complex 16 on September 20, 1984.

THOR

The Thor IRBM weapons test program had a short and fast-paced tenure at Cape Canaveral during the urgent period of the Cold War in the late 1950's. A total of 49 missiles were flown to test the operational suitability of the vehicle in a 37 month period. First phase was to fly 18 research and development (R&D) missiles from January 25, 1957 to August 6, 1958. Eleven of these vehicles carried the AC Spark Plug guidance system and five flew with the GE nose cone to test the warhead system. Thirty-one missiles were allocated to the Initial Operational Capability (IOC) phase which lasted from November 5, 1958 to February 19, 1960. The launch facilities consisted of the large concrete hard stands at LC-17A and B and a tactical version installed at LC-18B. There were 5 Thor launches from pad 17A, 27 from pad 17B and 17 launches from the tactical pad at LC-18B.

A common sight in the testing years: a 65 foot Thor rocket at LC-17B being prepared for flight in the Weapon System 315A program, destined to eventually deploy 60 missiles in England as part of Operation Emily.

(Right) Hanger M was used to prepare Thors for flight after arrival from Douglas Aircraft Company in Santa Monica, California.

(Left) An early Thor missile arrives at the gantry for lifting and placement on the launch pad.

(Right) Thor 104 ignites for the 4th R&D test and the first from LC-17A. Missile broke up some 92 seconds into flight on August 30, 1957.

(Left) Thor 120 was the first to fly with the Mark 2 simulated warhead on February 28, 1958 from the B pad at Launch Complex 17.

(Below) The nose of Thor 104 protrudes from the penthouse roof at the top of the pad 17B gantry prior to a static firing test in August 1957. Note the aerodynamic spike at the top of the missile that serves as an angle-of-attack indicator.

(Left) Thor 146 is late into its countdown on December 16, 1958 at LC-17(test 2502). The successful flight marked the first dual Thor launch as a SAC crew launched Thor 151 from Vandenberg AFB as a CTL demonstration for RAF crews beginning their tests in April 1959.

(Right) Thor 187 prelaunch on May 12, 1959 at LC-17B. Typical of many Thor images there are pad crew milling about the fully loaded missile.

(Below) Launch of Thor 187 on a successful flight. A movie camera recorded the separation of the nose cone from the missile and was later recovered downrange from the Atlantic Ocean.

(Right) Ninth in the IOC phase, Thor 176 undergoes liquid oxygen loading prior to a successful flight from LC-17B on April 23, 1959. Test 763

Thor Tactical Pad:
(Left) Thor 230 in the LC-18B gantry at twilight. Demonstrating program maturity, the IRBM made a successful flight into the target area on October 28, 1959.

(Below) Thor 184 is ready for 10th IOC launch from pad 18B on May 22, 1959 (test 1003). Douglas designated this missile type as DM-18A.

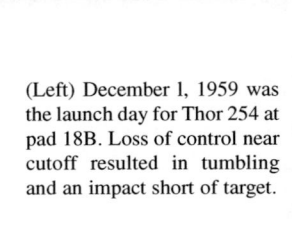

(Left) December 1, 1959 was the launch day for Thor 254 at pad 18B. Loss of control near cutoff resulted in tumbling and an impact short of target.

Nearing the end: Thor 255 thunders aloft from LC-17B on December 17, 1959 on the last IRBM shot from that pad. The successful flight was the 28th IOC test out of 31 conducted

JUPITER IRBM at Complex 26 and 5/6
The Army's Jupiter intermediate range ballistic missile was developed by Wernher von Braun's "Rocket Team" at the Redstone Arsenal in Huntsville, Alabama. The first example (AM-IA) was launched unsuccessfully from Complex 5 on March 1, 1957. Jupiter testing also took place at adjacent Complex 26 (right next door to the Air Force's Complex 17 for Thor) beginning in 1958. A total of 34 Jupiter launches took place at the two complexes including 5 Combat Test Launches for the Air Force (and Italian and Turkish allies that operated the missiles overseas).

A 60 foot Jupiter missile at Complex 26 in late December 1957. The code letters AM-H identify the vehicle as Army Missile number 4, although this was in fact the 7th test flight. The basic service structure at Complex 26 was compatible with both Jupiter and Redstone missiles.

(Left) Jupiter missile AM-30 (code TR) is prepared for the final research and development flight of the Jupiter program from pad 6 on February 4, 1960. The trailers hold the prelaunch checkout equipment.

(Right) Jupiter AM-26 (code AI) is nearly hidden within the unique gantry at Complex 26 prior to launch on December 16, 1959. The attached structure at the foot of the gantry enclosed work spaces for the launch crew. Note the photographer's tower in the foreground.

(Left) The moment of ignition for Jupiter AM-30 (TR) as the cooling and umbilical mast falls away.

(Right) The Cape scrub lands surround pad 26B at the Vertical Launch Facility as Jupiter missile AM-4 (code H) is prepared for launch by the Army on December 18, 1957. Note the ground support trailers that were used to house equipment used during the prelaunch checkout process.

(Left) Jupiter missile AM-23 is launched in daylight on September 16, 1959 from pad 26B. Within seconds of liftoff something went terribly wrong in the engine compartment and AM-23 suddenly veered off course. To see the final outcome of this launch, go to the Incidents and Accidents section.

(Right) Operational version of the Jupiter missile in USAF livery is lowered to the transport trailer in Hangar R in preparation for the first NATO training launch from pad 26A on April 22, 1961.

(Left) Jupiter missile 209 arrives at Complex 26 on March 6, 1961 for the first NATO combat training launch in April 1961 (test 1263).

(Right) Jupiter missile 209 is undergoing prelaunch checkout at pad 26A five days before the first NATO combat training launch. The Ground Support Equipment (GSE) is configured exactly as it would be at the overseas deployment bases in Italy and Turkey.

(Left) Combat training launch of a Jupiter missile from pad 26A. The "petals" on the ground around the rocket are part of a segmented environmental shelter used at the overseas bases.

The Flight of Able and Baker: (Right) The first living creatures to survive a rocket ride from Cape Canaveral were the female astro-monkeys Able and Baker, launched aboard Jupiter missile AM-18 (code PD) on May 28, 1959. Liftoff from pad 26B was at 0235 am EST (test 1751).

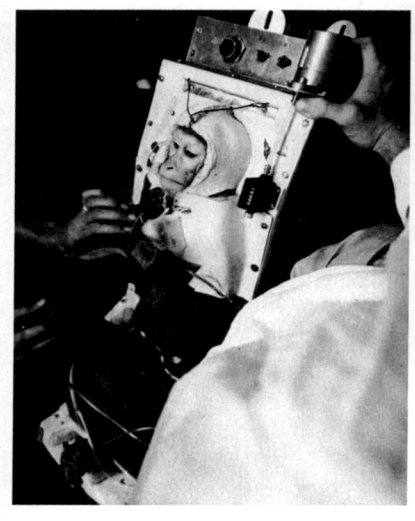

(Left) Monkey Able is secured in her form-fitting couch that fit inside a pressurized container in the large Jupiter nose cone. The second monkey, Miss Baker, was housed in a separate container.

(Right) The monkey container for Able is prepared for Bioflight-2 in Hangar R at Cape Canaveral. The container is presently preserved at the National Air and Space Museum in Washington, DC. The monkeys survived the 16 minute ride into space, but Able died during an operation to remove an implanted electrode. Miss Baker survived for 25 years at Pensacola, FL and the Alabama Space and Rocket Center in Huntsville, where she died in November 1984. Surviving a rocket ride landed the simian celebrities on the cover of LIFE magazine.

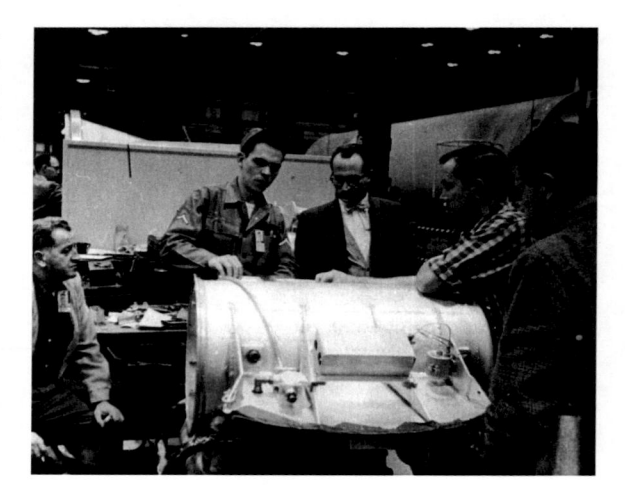

POLARIS FTV

Polaris Flight Test Vehicle (FTV) was a sub-scale prototype of Lockheed's submarine launched ballistic missile which utilized the XM-20 first stage motor of the X-17 missile. FTV was designed to test the configuration of the Polaris nose cone, and to test the operation of the Polaris "jetavator" steering system (moveable rings on each of the four first stage nozzles that steered the rocket during ascent). Early versions of FTV had boxy fins (later removed) and the final versions had odd flared skirts at the base of the missile. Seventeen FTV launches were conducted from pad 3 at Cape Canaveral from April 1957 to January 1958. Four additional FTV firings were made from pad 25A (April through June 1958) prior to the maiden launch of Polaris AX-1 in September 1958. All were apparently successful.

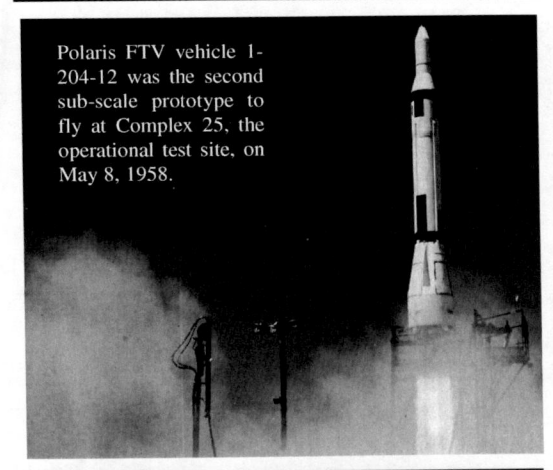

Polaris FTV vehicle 1-204-12 was the second sub-scale prototype to fly at Complex 25, the operational test site, on May 8, 1958.

POLARIS

Polaris was a new type of strategic weapon that was launched from beneath the sea by a nuclear submarine. The two-stage solid propellant missile was developed by Lockheed in the late 1950's to constitute the U.S. Navy's contribution to the American nuclear deterrent forces that eventually became known as the Triad (Air Force bombers, land-based nuclear missiles and submarine-launched ballistic missiles). Land-based testing of Polaris commenced at Cape Canaveral on September 24, 1958 from pad 25A. The Navy planned an extensive series of test flights to learn the art of solid propellant rocketry and to perfect the submarine launch technique, culminating in the first operational (nuclear-armed) cruise of the Polaris weapon system in November 1960. Development was so compressed that the first submarine test launch was conducted on July 20, 1960 from the USS George Washington, less than four months before the first operational cruise. The range of the initial Polaris A-1 model was 1,200 miles, and Polaris A-3 could reach 2,500 nautical miles. Polaris A-3 was first deployed in September 1964 (joining Polaris A-1 and A-2), an incredible logistical achievement for the Navy.

The first full-scale Polaris test missile (AX-1) rests on the metal legs of the launch mount at pad 25A prior to launch on September 24, 1958.

(Right) Sporting a gaudy pattern of tracking stripes, a Polaris A-3 missile (vehicle A3X-9) rises from pad 29A at Cape Canaveral on February 18, 1963.

(Below) Launch of Polaris A2X-2 vehicle at pad 25A on December 5, 1960.

(Inset, right) Polaris A-3 missile on the transporter-erector vehicle as it arrives at Complex 29.

(Right) The first Polaris A-3 prototype (A3X-1) is in position for launch at pad 29A, the second Polaris test site, in August 1962.

POLARIS SHIPBOARD LAUNCHES FROM USS OBSERVATION ISLAND

In addition to land pad launches of Polaris prototype missiles, the U.S. Navy arranged to test the Polaris submarine missile launch system in the late 1950's from a specially configured surface ship, the USS Observation Island (EAG-154) that was homeported during the test firings at Port Canaveral, Florida. The Navy selected a former Maritime Administration cargo vessel, the Empire State Mariner, to be fitted with a Polaris submarine launch tube. The six year old vessel was commissioned the USS Observation Island in December 1958. After sea trials the ship performed its first Polaris missile firing on August 27, 1959 in the waters off Cape Canaveral. After installation of the prototype Polaris submarine inertial guidance system (SINS), the Observation Island conducted six more Polaris launches in 1960, followed by test launches of the new Polaris A-2 model in 1961 and the A-3 model in 1963. The original Polaris launch tubes used compressed air to eject the missile, and the ship was re-equipped with a new steam ejection system to support Polaris A-3 launches in 1963. In 1968 the ship was converted to conduct test launches of the Poseidon missile beginning in 1970. The Observation Island was later converted to carry the COBRA JUDY missile tracking radar for the Ballistic Missile Defense Organization in the 1990's. The Observation Island was 564 feet long and displaced 12,978 tons. It had a complement of 434 men. The ship fired a total of 25 Polaris missiles and 3 Poseidon missiles from 1959 to 1970.

The Observation Island at sea with the USS Robert E. Lee (SSBN-601) in support of submerged Polaris launch from the submarine in the early 1960's.

(Left) A Polaris A-3 missile is ejected from the missile-firing tube of the USS Observation Island during a test flight off shore from Cape Canaveral.

CHEVALINE

Chevaline was the British Ministry of Defence's code name for an upgraded version of their Polaris A-3 submarine-launched ballistic missiles. The British Royal Navy accepted their country's nuclear deterrent role when the Skybolt air-launched nuclear missile program that was offered to the British was cancelled by the United States government in 1962. The Chevaline A-3TK system, which was designed to defeat Soviet anti-missile defense systems, was tested by the Royal Navy in a series of 11 flat pad flights at Complex 29 at Cape Canaveral from September 1977 to May 1980. Complex 29 was reactivated by the U.S. Navy for the Chevaline tests after being maintained in standby status since 1968. The British also test launched Chevaline from the Royal Navy's four nuclear powered missile submarines positioned off shore from Cape Canaveral. Chevaline was replaced by the much larger Trident II in Royal Navy service during the 1990's.

(Left) The seventh in the UK series of Chevaline land pad launches lifts off from pad 29A on September 1, 1979. The vehicle malfunctioned at 90 seconds and was destroyed by Range safety.

POSEIDON

Poseidon (C-3) was the Navy's replacement for the submarine-launched Polaris A-3 missile deployed in Fleet Ballistic Missile boats since 1964. Poseidon was 20 inches wider and 3 feet longer than Polaris A-3, but could fit in the same launch tubes in the submarines. The missile had the same 2,800 mile range as Polaris A-3, but featured improved solid motors that could carry a greater offensive payload (throw-weight). The first land pad launch of Poseidon from pad 25C took place on August 16, 1968 (coincidentally the same day as the maiden launch of the Minuteman III from neighboring silo 32B). There were sixteen land launches of Poseidon before testing transitioned to submarines on August 3, 1970 (from the USS James Madison). Poseidon was finally retired from service in 1994 after being replaced by the more powerful Trident I (C-4) missile.

(Above) Liftoff of the 9th Poseidon test missile (C3XT-11) from pad 25C at Cape Canaveral on July 9, 1969. Note the display missiles of the museum on the horizon at left.

(Right) Remarkable close up of Poseidon C3XT-21 as it lifted off from pad 25C on May 14, 1970.

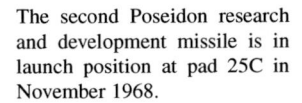

The second Poseidon research and development missile is in launch position at pad 25C in November 1968.

TRIDENT I
The Navy's Trident I (C-4) was the first missile to be deployed aboard the enormous 560 foot Ohio-class missile submarines ('boomers') with 24 launch tubes that began deterrent patrols in 1980. The 34 foot long Trident I (74 inch diameter) was substantially larger, with greater range, than its Poseidon predecessor. Trident I testing got under way from pad 25C at Cape Canaveral on January 18, 1977. The 17th test launch on December 17, 1978 veered off course immediately after liftoff and was destroyed by the RSO after only 5 seconds aloft, one of the most spectacular accidents in the history of the Cape (see Incidents and Accidents). Trident I was withdrawn from service in 2000-2003 in favor of the much larger Trident II (D-5) missile. The final launch of a Trident I took place on December 18, 2001 from the USS Ohio, one of four missiles launched that day off-shore from Cape Canaveral.

(Right) The 18th and final Trident I land pad launch occurred on January 23, 1979 from pad 25C at the Cape.

(Left) An early version of the Trident I missile lifts off from pad 25C. The missile's aerospike on the nose has already deployed.

TRIDENT II
Trident II (D-5) is the largest submarine-launched missile ever to be deployed by the U.S. Navy, carried in the 24 launch tubes of the Ohio-class submarines since the 1990s. The three stage solid missile is 44 feet tall and measures 83 inches in diameter. With its 10,000 Mile range, the weapon was denounced as a first-strike weapon by the former Soviet Union. With the retirement of Trident I in 2003, Trident II became the only submarine launched ballistic missile in the U.S. Navy inventory. The two-year development test program for Trident II began at Complex 46 on January 15, 1987. Nineteen test launches were conducted before sea trials began on March 21, 1989 from the USS Tennessee. The first sea launch was a spectacular failure: the missile did a loop-de-loop immediately after exiting the surface (see Incidents and Accidents) before being destroyed by range safety.

(Left) The Navy's fifth Trident II test missile leaps off the launch pad at Complex 46 on July 20, 1987. The wires strung between two poles provide protection from Florida's frequent lightning strikes.

THE NAVAL ORDNANCE TEST UNIT (NOTU)
The U.S. Navy's presence at Cape Canaveral is represented by the Naval Ordnance Test Unit (NOTU), headquartered on Pier Road near the main gate of the base. The Navy has been represented at the Cape since 1950 when the Joint Long Range Proving Ground was created, and NOTU was organized in 1956 to coordinate the increasing level of Navy activity at Cape Canaveral (primarily at the port). The Navy was preparing for the coming of sea-launched ballistic missile testing to Canaveral (initially intended to be ship-launched Jupiter missiles, until the Navy chose solid propellant Polaris missiles launched from submarines to form their nuclear deterrent force). NOTU's primary function remains to support SLBM testing (sea-launched Trident II missiles) along with maintaining the port facilities. NOTU homeports a Launch Area Support Ship (LASS) at Port Canaveral, the USNS Waters, which accompanies each submarine to sea for the test launches.

(Left) The entrance to NOTU headquarters on Pier Road near the south gate of Cape Canaveral.

(Right) Sinister black shape of a Trident nuclear submarine at the Navy Wharf on the Trident Basin at Port Canaveral.

SEA LAUNCHES
By far the vast majority of the Navy's submarine-launched ballistic missile tests since 1958 have been performed off shore from Cape Canaveral by submerged subs (over 900 launches). The submarines sortie from the wharf of the Naval Ordnance Test Unit (NOTU) at Port Canaveral (on the southern boundary of Cape Canaveral, near the main entrance to the base). The facility is now called the Trident Basin after the channel was dredged for the massive Trident missile-carrying subs in the mid 1970's. The missile boats (called "boomers" in the Navy) sail off shore to a classified firing zone to conduct the tests, accompanied by the Launch Area Support Ship, the USNS Waters. The subs only fire one or two Trident missiles in present day tests, but as many as six or seven Polaris A-3 missiles were fired in one session during the height of the Cold War in the 1970's. Lockheed Martin currently boasts of 111 successful Trident II sea-launches in a row through September 2005.

(Right) The temporary telemetry mast used for sea-launched missile test flights is about to be installed on the sail of the USS Thomas Jefferson at Port Canaveral in February 1963.

(Right) One of the first Polaris A1-E1 missiles is launched from a submarine off Cape Canaveral in 1960.

(Below) An orbital perspective on a U.S. Navy Polaris sea launch: the Gemini 7 astronauts managed to capture this image of a Polaris missile in flight (the streak with a bright tip, top right) on December 6, 1965.

(Right) A Royal Navy Polaris A-3 missile is successfully launched from HMS Revenge off Cape Canaveral on July 25, 1986.

(Right) The 57th submerged launch of a Poseidon C-3 missile took place on July 24, 1984 from the USS George C. Marshall. Note the telemetry mast at right.

(Left) This Trident II (D-5) missile is already heading downrange after surfacing off the east coast of Florida on December 4, 1989. This fourth Trident II sea launch was conducted from the USS Tennessee.

ATLAS

America's first long range missile was tested extensively from 1957-1962 from four launch pads at Cape Canaveral. A total of 83 vehicles (and 6 different models) were tested encompassing multiple configurations, ranges and re-entry vehicles. Launch Complexes 12 and 14 supported the eight A series missiles while the B model (closely resembling the operational missile) inaugurated LC-12 and LC-13 for a nine missile test series. Follow-on C and D series brought the Atlas to operational status during 38 test firings from 1958-1961. The more advanced E and F versions were qualified by a final 18 missile test series from 1960-1962. Ranges flown were from 600 miles to 9,000 miles down the Atlantic Missile Range with four different re-entry vehicles. Atlas also became a space launch vehicle during its test period at the Cape and was used with many upper stage configurations including Able, Agena and Centaur from 1959- 2005.

(Left) The second Atlas flight article, 6A, gleams in the sunlight on pad 14 during preflight checkout in September 1957. The gantry support rails are visible at the base of the pad. Note the technical support trailers parked on the launch pad.

(Below) Atlas 4A at ignition on June 11, 1957 at LC-14 (test 0895). After a good liftoff at 2:37 pm and pitchover downrange, excessive heat in the engine compartment caused both 135,000 pound thrust chambers to shut down. Atlas then proved the structural integrity of the airframe by executing several violent pitch maneuvers prior to range safety destruction less than one minute into the ascent.

(Left) First Atlas to fly from LC-12 and to use the familiar 1,000 pound thrust vernier engines, 10A is seen under the portable searchlights in late November 1957 during preflight testing. On January 10, 1958 Atlas 10A made the second successful Atlas flight in four attempts.

(Right) A Convair technician inspects the wicked-looking aerospike mounted on the nose of an Atlas A missile (serving as an angle-of-attack indicator) at Cape Canaveral hangar.

(Left) The first flight Atlas (4A) is a dominating presence on pad 14 during the pre-launch checkouts conducted in May and June 1957.

(Right) Atlas 13A is seen being backed up to the launcher arm assembly, after which it was attached and was rotated upright by a cable extended from the top of the gantry. 13A flew an unsuccessful R&D mission February 7, 1958 from LC-14.

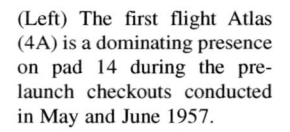

The Atlas B series was the first to be equipped with the full Atlas propulsion system. Atlas 5B is seen from the LC-11 perimeter road looking to the south. This was the second and last B model to carry the black and white paint pattern. 5B was delayed by some on-site welding repairs and X-ray tests up in the gantry, which saved Convair the time and effort of removing the missile from the pad. Shortly before midnight on August 28, 1958 Atlas 5B made a successful flight and separated its Mark 2 nose cone.

(Right) Atlas 12B was the seventh test flight of the series and was part of the Hotrod series (6B, 10B, 12B) that were allocated the best-performing engines from Rocketdyne to attempt long distance flights. 6B made the first attempt to go full range in September 1958, but it suffered a sustainer engine failure. On the evening of November 28, 1958 Atlas 12B earned the honor of a 5,500 statute mile flight validating the Atlas design.

The first of the flight rated D series was 3D flown unsuccessfully from LC-13 on April 14, 1959. This missile was very close to the operational version expected to be deployed later in the year. Two major ground support malfunctions at liftoff caused an erratic flight and sub-sequent destruction by Range Safety at 36 seconds (see Incidents and Accidents).

(Right) Atlas 5D is framed by the many searchlights used for launch photo coverage. The failure of 5D in flight was the fifth straight and caused the USAF to order a stand down just as the Atlas was moving towards being declared operational at Vandenberg AFB.

(Left) The power and fury of Atlas 55D's propulsion system is only hinted at as the 3 main engines and the 2 verniers ignite on October 22, 1960 for a successful R&D flight far out into the Atlantic Ocean. The 12:13 AM launch sent the Mark 3 RV to a target area 6300 nautical miles away. This was the last ICBM flight from LC-14.

(Right) Looking up the ramp at the departure of Atlas 76D on a successful 4,000 mile flight in the early evening of September 16, 1960. This missile was one of the D models allocated to carry guidance system components for the upcoming all-inertial Atlas E and F models. D models were designed with a radio guided system and had a much smaller equipment pod set mounted in line with the pitch axis just above the engine compartment. The large RVX-2 experimental nose cone was not recovered from the ocean.

(Left) Workman on a ladder at pad 12 makes an adjustment to the fairing of the downrange antenna. Atlas 90D flew a successful test flight down the Atlantic Missile Range on January 23, 1961 (test 3505).

(Below) Atlas 90D strikes a classic pose at LC-12, loxing in the late afternoon sun on January 23, 1961 just before the 32nd and final D series R&D test at the Cape.

(Left) The first E version to fly was Atlas 3E from LC-13 on October 11, 1960. This test carried the larger and heavier Mark 4 RV as designed for the Atlas E/F and Titan I ICBMs. Note the lack of hold down arms; the E and F series were designed to ignite the engines and lift off when thrust reached the necessary level. The umbilical mast with its boom is not being used for connections as the necessary linkups are done through the base of the missile. A hydraulics failure led to the break-up of 3E at 154 seconds into flight.

Convair technicians make final connections with the new hold-down mechanism at pad 13 that duplicated the support structure of the Atlas E semi-hardened shelters deployed in the mid-western United States. Atlas 4E was the second test vehicle in the series launched from Cape Canaveral on November 30, 1960.

(Left) Atlas 8F is trucked past the blockhouse to Launch Complex 11 for the eighth F model R&D test. A Mark 4 RV was the payload, and a Scientific Passenger Pod (SPP) was also carried on the successful flight on September 19, 1962. The F-model was deployed in silos at six Air Force bases beginning in 1962, with 12 nuclear armed missiles at each base.

Last of the 83 research and development Atlas ICBM test missiles: 21F is prepared for launch on the late afternoon of December 5,1962 on a full-range shot down the Atlantic Missile Range. Atlas missiles were operational from 1959 to 1965 during the Cold War and were a major deterrent to the Soviet military threat.

TITAN I

The second USAF ICBM was the two-stage Titan I missile built by Martin, which was radio-guided to its target. An initial R&D series of 4 live first-stage only vehicles was successfully flown in the spring of 1959, followed by the early 2 stage versions which flew limited range flights. With the introduction of the Mark 4 re-entry vehicle, the flights were extended to full range and successes eventually became routine (there was significant trouble with the upper stage in the initial firings). The USAF used the northern end of ICBM Row for the test program which expended 47 missiles over a 36 month period, indicative of the high priority accorded ballistic missile testing in the cold war era. Titan I was 98 feet tall with its RV. A new concept of launch processing was developed by Martin for Titan I: a hydraulically actuated erector gantry which remained at the pad and did not require a tower rollback as was the case for other missiles at the Cape. Titan I R&D variants were designated by lettered Lot numbers (Lot A, Lot B, Lot C, Lot G, Lot J, Lot AJ and Lot M).

The gantry at LC-20 is being lowered on July 1, 1960 in preparation for the first launch from the new complex with Titan I vehicle J-2 (test 1801). A broken hydraulic line in the first stage created a spectacular off-course flight of 11 seconds with both stages being destroyed by Range Safety, dumping large loads of burning propellants and debris within 2,500 ft. of the Titan launch pads at Cape Canaveral.

A new generation of U.S. ICBM's is erected at LC-15 and readied for testing. Titan I-A-3 was launched on a live first stage/dummy second stage flight on February 6, 1959. Note the reinforced fabric "curtains" used on the upper levels of the gantry to provide modest protection from the elements.

The fourth and last Lot A Titan I missile is shown in the erector gantry at LC-15, prior to a successful shot on May 4, 1959 to round out a 4-for-4 launch record during the first half of 1959. The immediate future would not be so kind, with failure of the first missile equipped with a live upper stage.

(Left) A-4, the second Titan I flight vehicle, stands erect on LC-15 with the erector gantry in the process of retraction to the parked (ground) position. Liftoff was on April 3. 1959. A second firing position at left was provided for static testing of the Titan second stage.

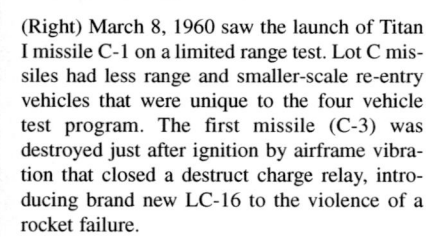

(Right) March 8, 1960 saw the launch of Titan I missile C-1 on a limited range test. Lot C missiles had less range and smaller-scale re-entry vehicles that were unique to the four vehicle test program. The first missile (C-3) was destroyed just after ignition by airframe vibration that closed a destruct charge relay, introducing brand new LC-16 to the violence of a rocket failure.

(Left) The first true Titan I long range missile (approximating the operational version) was G-4, which carried a separable Mark 4 RV on an intercontinental range flight on February 24, 1960 from LC-15. Titan would require another 23 months of testing using 36 missiles that would advance vehicle range and reliability.

(Right) Last of the lot C Titans(C-6 is enclosed in its gantry at LC-16 in this view looking south-east down ICBM Row in April 1960. LC-15 and two of the four Atlas gantries are visible. The launch on April 28 was successful.

J-13 is shown at LC-19 without its Mark 4 reentry vehicle. A successful 5,000 mile flight was made in February 1961 on the 29th R&D flight.

(Right) Vehicle AJ-12 lifts off from LC-20, the northernmost Titan I pad at Cape Canaveral, on a partly successful flight on March 2, 1961. The AJ series were nearly identical to the production line vehicles deployed at the Titan I bases and represented the final validation of the Titan I design.

(Below) An unidentified Titan is imaged by a remote controlled camera located in the reclining erector gantry. Oxygen tanks on both stages are covered with frost as the time of launch approaches. Titans at operational sites were launched from elevator platforms that brought the missile to the surface from the underground silo.

(Right) A view looking southeast at J-2 in the elevated gantry at LC-20. The Titan I gantries were more open and compact than the neighboring Atlas structures. Despite the hydraulics needed for movement the Titan service towers were probably less costly in maintenance.

(Above) A twilight launch of J-9 on December 20,1960 from LC-20 on the 26th R&D flight. The second stage failed to ignite and impacted in the Atlantic Ocean.

Symbolic of a power never to be used: a two stage Titan I with a simulated warhead in its Mark 4 re-entry vehicle ascends into the skies above Cape Canaveral. 54 of the 98 foot tall missiles were deployed in silos from 1962-1965 across the western United States.

Minuteman I, II, III

America's first generation ICBM's, Atlas and Titan I, were liquid fuelled rockets that were difficult to maintain and slow to react in times of crisis. The Air Force decided to develop a solid propellant missile (then a very new technology) to replace Atlas and Titan, and the Boeing Company was awarded a contract to provide the Minuteman ICBM (WS-133A) in October 1958. The missile was 58 feet long and 6 feet in diameter with a range of 6,300 statute miles. With a sense of urgency unknown in the present day, Boeing was ready to test fire the first Minuteman at Cape Canaveral after only 28 months development time. The first launch on February 1, 1961 was a complete success, marking the first time that a missile and its upper stages were tested together on a maiden flight, a process known as "all-up" testing. After four above-ground firings, Boeing proceeded to test fire the missile at Complexes 31 and 32 from 80 foot deep underground silos. The first launch ended in a fiery disaster, but subsequent tests validated the underground launch concept. Minuteman I and II carried a single Mark 11 re-entry vehicle. Boeing introduced an improved Minuteman II version in September 1964 that featured a larger second stage motor, and the more capable Minuteman III with an uprated third stage was launched for the first time on August 16, 1968. Minuteman III was designed to carry three independently targeted re-entry vehicles, and is still in service today (now limited to a single warhead) as America's only land-based nuclear missiles. Minuteman testing ceased at Cape Canaveral in December 1970 after a total of 92 missiles were launched. The first Minuteman missiles went on nuclear alert in December 1962 at Malmstrom AFB in Montana, and by June 1965 they filled 800 silos on the western plains of the United States.

(Right) Test 3058. The first Minuteman missile (number 401) stands ready for launch at pad 31 A in February 1961, an above-ground facility used for the first four Minuteman tests before the commencement of underground silo test launches.

(Left) Liftoff of Minuteman 401 on February 1, 1961 for a completely successful maiden flight, a rare occurrence in the early days of missile testing.

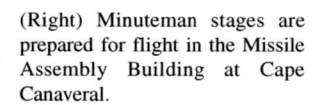

(Right) Minuteman stages are prepared for flight in the Missile Assembly Building at Cape Canaveral.

(Left) Fully assembled Minuteman I missile (except for the reentry vehicle) inside the Missile Assembly Building, ready for transport to the pad.

(Below) Minuteman I number 421B is slowly inserted into the transport canister in February 1963 before being installed in silo 32B.

(Left) An early Minuteman is levered into launch position on the flat pad at Complex 31. Missiles being installed in the underground silos were inserted directly from the transport canister (above).

(Right) Launch control team at their stations inside one of the distinctive beehive-shaped Minuteman block-houses at Complexes 31 and 32.

(Left) A Minuteman II missile installed inside one of the silos at Cape Canaveral.

(Right) Minuteman missile blasts out of its silo at the Cape in 1962.

(Left) Spectacular silo launch of a Minuteman II missile at Cape Canaveral, complete with the characteristic smoke ring at the top of the picture. The beehive-shaped blockhouse is in the foreground.

(Right) The final Minuteman III launch at Cape Kennedy on December 14, 1970 from Silo 32B. The shot was one of three conducted by Boeing to verify range and accuracy of the missile before the Air Force accepted the, weapon, designated the Special Test Missile Project.

TITAN II ICBM
The third and final liquid propelled ICBM developed by the U.S. Air Force was the Martin-built Titan II. The vehicles used storable hypergolic propellants that ignited on contact, which improved the missile's response time, and permitted extended periods of storage in the silos with propellants aboard the missile. The 103 foot missile was the largest ICBM ever developed in the United States, producing 430,000 pounds of thrust at liftoff. The payload was a massive Mark 6 nuclear warhead. The Air Force flew a 23 vehicle research and development program at Cape Canaveral from March 1962 to April 1964 to qualify Titan II for active duty at bases in the southern United States (the last operational Titan II was removed from its silo in 1987). Martin and the Air Force flew all of the R&D vehicles from LC-15 and 16, since the other two former Titan I pads (LC-19 and 20) were allocated for space missions. Fundamentally the Titan II evolved very little during the testing phase: from day one all the missiles had live second stages and were programmed to fly full-range missions. The Air Force worked in close coordination with NASA to adapt the Titan II missile for the manned Gemini space flights. During the test phase these two organizations managed to cure the inherent pogo vibrations in the Titan II propulsion system to qualify the booster to carry astronaut crews.

The first Titan II flight article was vehicle N-2, shown here at pad 15 prior to the maiden flight on March 16, 1962. N-2 flew a successful full-range mission, duplicating the first Minuteman I "all-up" flight in February 1961 as a sign of maturing missile technology.

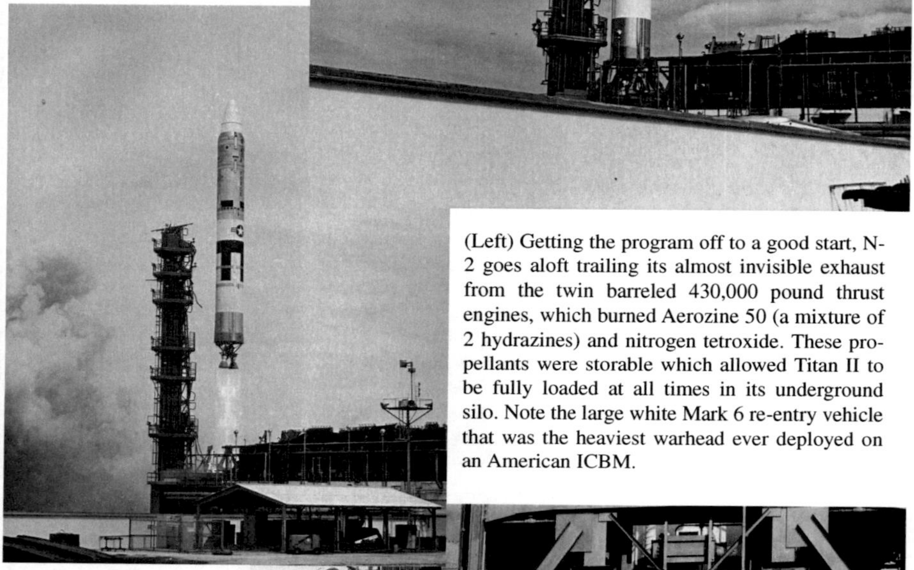

(Left) Getting the program off to a good start, N-2 goes aloft trailing its almost invisible exhaust from the twin barreled 430,000 pound thrust engines, which burned Aerozine 50 (a mixture of 2 hydrazines) and nitrogen tetroxide. These propellants were storable which allowed Titan II to be fully loaded at all times in its underground silo. Note the large white Mark 6 re-entry vehicle that was the heaviest warhead ever deployed on an American ICBM.

The launch mount for the Titan II missile is featured in this portrait of vehicle N-31 at pad 15 in January 1964. Two of the hold-down arms are visible, framed by the metal beams in front of the launch vehicle.

(Left) A unique test flight for Titan II was the unsuccessful launch of N-11 on December 6, 1962 carrying a Mark 4 re-entry nose cone in place of the normal Mark 6. N-11 is seen departing LC-16 in the late afternoon. Severe vibrations caused a second stage engine pressure switch to shut down the Aerojet engine prematurely.

(Above) A Titan II is being unloaded from a C-133 transport aircraft after a flight from Denver to participate in the 23 missile research and development program from 1962-1964 at Cape Canaveral.

(Left) Titan II vehicle N-21 is being stacked at pad 15 for flight in April 1963.

A forklift is used to carefully lift an interstage structure for mating to an R&D Titan II. The open blast ports, a unique feature on Titan, permitted the second stage to ignite while still attached to the first stage. This was called the "fire in the hole" staging technique where the second stage exhaust escapes from the openings in the interstage unit prior to stage separation.

(Below) The first Strategic Air Command crew to participate in a Titan II R&D launch successfully flew N-16 on February 16, 1963 from LC-15 on a full range test. These were called "blue suit crews" to distinguish them from the Martin contractor people and to denote them as USAF personnel.

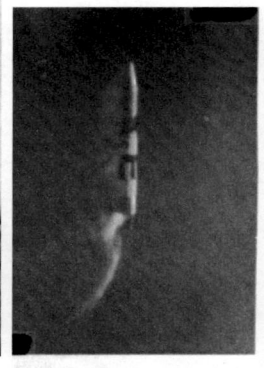

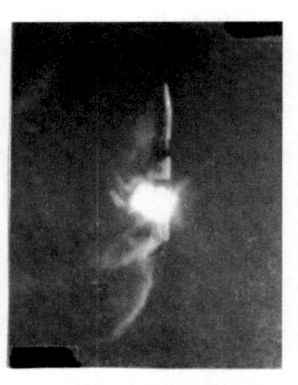

(Right) The destruction of vehicle N-20 on May 29, 1963 (test 124) is illustrated in these frames from the IGOR tracking telescope at Patrick AFB. A malfunctioning first stage activated the destruct system on N-20 50 seconds after liftoff (frame 2). The second stage was also destroyed by the RSO as a precaution (frame 4). N-20 was the only Titan II to be destroyed in flight, a sharp contrast to the failure records of older generation missiles at Cape Canaveral.

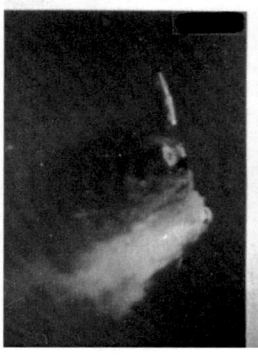

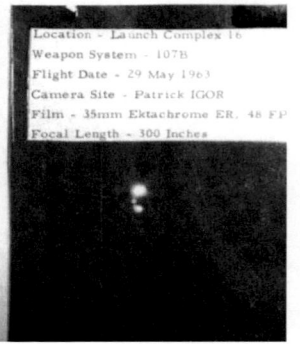

(Left) A Titan II in action at Launch Complex 15. Test 525 was the successful test flight of vehicle N-31 launched on January 16, 1964 as the Air Force and Martin strove to wrap up the Titan II development program by April of 1964.

(Right) View looking north of the Titan II pad at Complex 15 just off the vehicle access ramp where the erector gantry is lowered to the parked (ground) position.

RV-A-10
Thiokol's RV-A-10 experimental missile was the first solid propellant rocket flown at Cape Canaveral, and the largest solid rocket in the United States at the time. Developed as part of the General Electric Company's Project Hermes, the 20 foot RV-A-10 vehicle was flown four times from pad 3 at Cape Canaveral in February and March 1953. The 26 inch diameter "Glue Pot" motor had a burn time of 45 seconds, the longest duration solid motor of that time, capable of boosting RV-A-10 to an altitude of 58 km. The successful launches of this vehicle proved the feasibility of large grain solid propellant rockets.

General Electric Company diagrams of the RV-A-10 solid propellant test missile flown at Cape Canaveral in 1953.

Figure 30. Free-Flight Test Vehicle RV-A-10

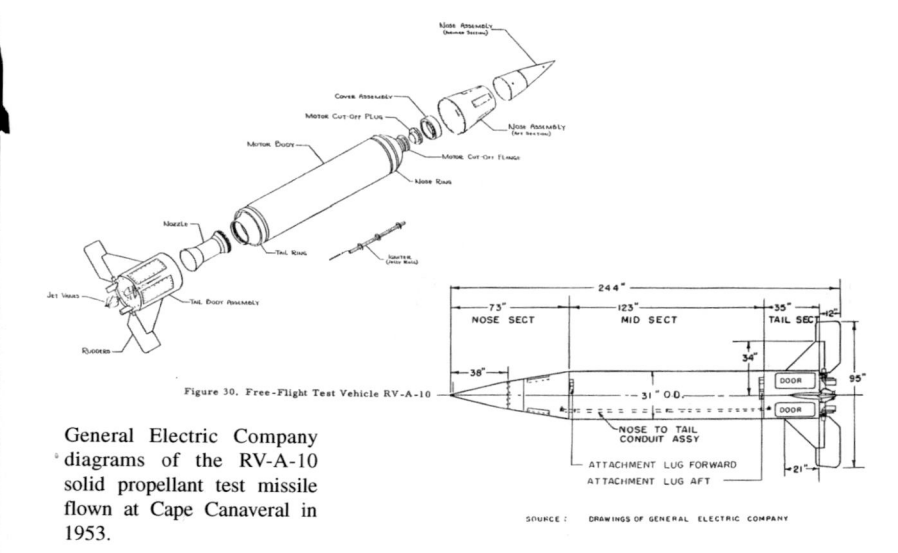

Figure 29. Propulsion Test Vehicle RV-A-10

X-17

Lockheed's X-17 was a three-stage solid fuel missile used to launch experimental reentry vehicles that provided crucial data for the Air Force Ballistic Missile Division's Thor and Atlas programs. The two upper stages (Recruit motors) of this 40 foot rocket were fired in a downwards direction to accelerate the model reentry bodies to a velocity of up to Mach 20. The first stage consisted of a Thiokol XM-20 Sergeant motor. Thirty eight X-17's were flown off of pad 3 (the Cape's original launch site) from August 1955 to December 1957. In addition there were six sub-scale X-17 prototypes that were launched beginning in May 1955 as proof-of-concept tests from pad 3. Large spin rockets on the first stage gave the X-17 its distinctive corkscrew motion at liftoff.

The final X-17 vehicle of the program (R-25) stands ready for launch at pad 3 on March 21, 1957. Note one of the original Cape blockhouses at left.

JASON

Jason (Argo E-5) was a five stage solid fuel suborbital research rocket launched from Cape Canaveral during 1958 in support of the three Project Argus high altitude nuclear tests conducted from the navy ship USS Norton Sound on the Atlantic Ocean. Six Jason rockets were fired from two sites at Canaveral, pad 4 (Bomarc site) and pad 10 (former Navaho tactical pad) to record data from the Argus shots on August 27, August 30 and September 6, 1958. Yield of the Argus devises, (warheads from the Genie missile) was 1.7 kilotons. Additional Jason rockets were also fired from Wallops Island, Virginia and Ramey AFB in Puerto Rico to observe the Argus shots.

(Left) The first Jason sounding rocket is prepared for launch at Complex 10 to measure radiation from the Project Argus atomic test in August 1958.

(Right) U.S. Navy photograph of one of the three X-17 rockets for Project Argus, ready for launch from the USS Norton Sound on the Atlantic Ocean.

JUPITER-C

In 1956 the U.S. Army originated development of the Jupiter-C ('Composite') re-entry test vehicle by adding a cluster of solid propellant rocket motors to the top of the existing Redstone rocket. The Jet Propulsion Laboratory designed a spin-stabilized cylindrical tub to house the 14 Baby Sergeant motors - 11 motors arranged in an annular pattern surrounding three motors in the center. The project was intended to test a scale model of an advanced ablative nose cone the Army was developing for the Jupiter IRBM. Three flights were conducted between September 20, 1956 and August 8, 1957 in the early morning darkness so downrange forces could see the re-entry prior to the recovery attempt. The solid rocket motors propelled the RV to an altitude in excess of 600 miles, to achieve maximum velocity for the descending sub-scale nose cone. The first flight was a test of the entire system with a dummy RV. Of the two re-entry test flights, only the second RV was recovered. Leftover vehicles were put into storage by the Army at Huntsville for the day when the re-named Juno I vehicle would emerge into the spotlight, to launch Earth orbiting satellites with the addition of a fourth stage motor to the Jupiter-C configuration.

A dummy fairing was installed in place of the 1/3 scale Jupiter nose cone on the first flight in September 1956 (left), from Launch Complex 5 at Cape Canaveral.

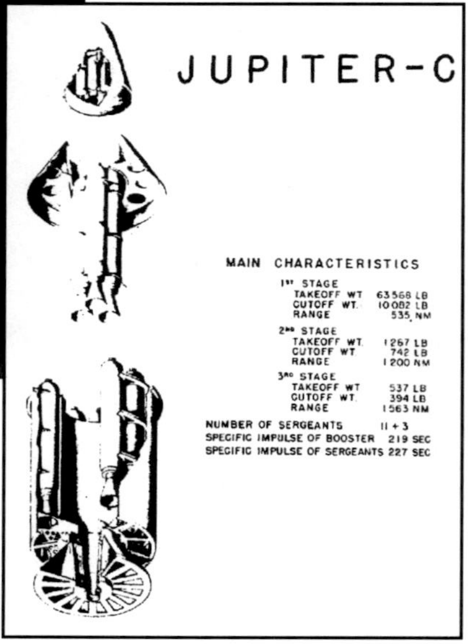

The first Jupiter-C vehicle was only a test of the 'Composite' solid rocket cluster, at right.

JUPITER-C

MAIN CHARACTERISTICS

1ST STAGE	
TAKEOFF WT	63568 LB
CUTOFF WT	10082 LB
RANGE	535 NM
2ND STAGE	
TAKEOFF WT	1267 LB
CUTOFF WT	742 LB
RANGE	1200 NM
3RD STAGE	
TAKEOFF WT	537 LB
CUTOFF WT	394 LB
RANGE	1563 NM
NUMBER OF SERGEANTS	11 + 3
SPECIFIC IMPULSE OF BOOSTER	219 SEC
SPECIFIC IMPULSE OF SERGEANTS	227 SEC

Images of the Jupiter-C are rare in the public domain; the 3 images in this section are of the first test vehicle, Redstone number 27 (code UI) for the September 20, 1956 launch. The cylindrical tub above the Red-stone nose was spun by electric motors so as to minimize any solid motor exhaust deviations.

The gantry is partially retracted in this view of a pre-flight test as noted by the lack of liquid oxygen frosting and venting on the lower part of the tank structure.

BLUE SCOUT JUNIOR
Blue Scout Junior was a spin stabilized solid propellant research rocket that was capable of carrying a small payload of scientific instruments for the Air Force to extremely high altitudes (in excess of 25,000 miles in some cases). Originally designated the Hyper Environmental Test System - 609A (HETS program 609A), Blue Scout Jr. (SLV-1B) was used to investigate the Van Allen radiation belts during ten flights at pad 18A from 1960-65. The diminutive four-stage rocket was 40 feet long.

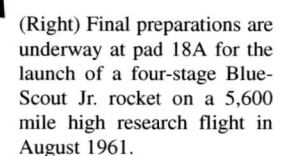

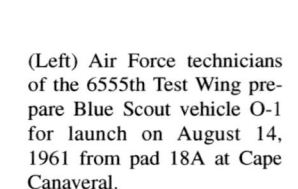

(Right) Final preparations are underway at pad 18A for the launch of a four-stage Blue-Scout Jr. rocket on a 5,600 mile high research flight in August 1961.

(Left) Air Force technicians of the 6555th Test Wing prepare Blue Scout vehicle O-1 for launch on August 14, 1961 from pad 18A at Cape Canaveral.

(Right) The second Blue Scout Jr. rocket was fired from pad 18A at Cape Canaveral, former Vanguard site, in November 1960.

Blue Scout Jr. rocket 22/1 was launched in July 1963 to study astronomical radio waves for Harvard University (top and bottom).

BLUE SCOUT I, II

Blue Scout I and II were solid fuelled suborbital rockets flown by the Air Force from Complex 18B in 1961 and 1962. Blue Scout I was a three-stage rocket based on NASA's Scout launch vehicle, and Blue Scout II (with four stages) was identical to the NASA Scout. Only five of these vehicles were flown before the program was cancelled in 1962, after the ASSET program was transferred from Blue Scout to Thor. The program was plagued by two launch failures and an inability to recover the small reentry capsules carried on several flights. Aeronutronic Ford was the Blue Scout contractor for the Air Force (including Blue Scout Junior).

(Right) Blue Scout I vehicle at pad 18B.

(Below) Night launch of a Blue Scout II from the same facility.

(Right) A Blue Scout rocket is emplaced at pad 18B. The top half of the launch table is already attached to the rocket as it is levered into position by the strong arm of the transport vehicle.

(Left) The first Blue Scout I vehicle (D-3) on pad 18B prior to launch on January 7, 1961 (test 2510). Blue Scout utilized the existing gantry at pad 18B, inherited from the tactical Thor program.

WEATHER ROCKETS AT LC-43 AND LC-47

Meteorologists have long utilized weather balloons and small rockets to improve their local forecasts. The U.S. Air Force supported a network of weather rocket launch sites around the world since the 1960's, and one of these sites was complex 43 at Cape Canaveral. According to the 45th Space Wing, more than 4,680 small weather rockets (Loki, Arcas, Judi-Dart and more) were fired from the launch rails at complex 43 from 1962 to 1984. In March of 1984 the weather launches were transferred to a new site designated complex 47, when construction for the new Navy launch pad for Trident II began at the former location. Pad 47 has also been used for occasional student experiments using Super Loki rockets. The Florida Space Authority currently operates pad 47 through a 'no cost' licensing agreement with the Air Force in effect since June 2003.

(Left) A Robin weather rocket is fired from the Arcas launcher at Cape Canaveral on October 2, 1959 within sight of the Cape lighthouse and Hangar C.

A slender Loki-Dart weather rocket is launched from Complex 43 on June 26, 1970.

SOUNDING ROCKETS AT CAPE CANAVERAL

In addition to numerous small weather rockets launched from Complex 43 from 1962 to 1984, a lesser number of scientific sounding rocket launches were also conducted from pad 43 in the 1960's. In 1963 the Air Force conducted preliminary atmospheric measurements with six Nike Cajun sounding rockets for Project TRUMP - Target Radiation Measurement Program. The Air Force later launched six Nike Tomahawk rockets in January 1967 from pad 43 to study upper atmospheric conditions in the MUMP project (Marshall - University of Michigan Probes). NASA also launched over 100 single-stage Nike Smoke rockets to trace upper atmosphere wind patterns in the mid 1960's. The only other sounding rockets launched from Cape Canaveral were the Jason rockets in 1958, and several large military research rockets from Complex 20 in the 1990's.

Two views of a Nike Cajun sounding rocket set up at Complex 43 (site D) for Project TRUMP in early 1963.

ALPHA DRACO

Alpha Draco was McDonnell Aircraft's entry into an Air Force-sponsored competition in 1959 to develop an air-launched, long range ballistic missile (ALBM). McDonnell chose to test fly a special lifting body reentry vehicle intended for their ALBM entry. The Air Force allocated the old Navaho tactical pad (complex 10) to the project, and McDonnell built a simple concrete flame deflector for Draco at the site. The 46 foot long, solid fuel missile was launched three times at pad 10 from February to April 1959. Two flights were successful, but the competition was won by Douglas, who went on to build the short-lived Skybolt missile. In 2000 the author helped to rediscover the Draco launch emplacement at pad 10 along with two Missile and Space Museum volunteers.

The third Alpha Draco vehicle is prepared for launch at pad 10 in April 1959. McDonnell utilized an Honest John transport vehicle as the Draco launch mount. Draco was fired at a 100 degree true azimuth angle that deposited the vehicle in the Atlantic Ocean. No recovery attempts were planned.

(Below left and right) Draco is secured for launch above the concrete flame deflector at pad 10.

(Below and Right) The Draco launch site today. The clearing in the upper part of the picture is where Draco was parked for the launches in 1959. The fire hydrant in the bushes (lower left corner) is the same hydrant visible in the photo on the previous page, which confirms the area as the Draco launch site. (Right) The edges of the concrete flame deflector for Draco are visible in this photo taken by the author at old pad 10 in August 2000.

(Right) The third Draco vehicle streaks aloft from pad 10 on April 27, 1959.

Distant shot of the third Draco launch taken from the perimeter road of Complex 17. The recently abandoned Navaho pad is at left and the still-active Vanguard site (pad 18A) is visible to the left of the Draco in flight.

BOLD ORION AND HIGH VIRGO
Bold Orion (pictured) and High Virgo were the other two contestants in the Air Force's air-launched ballistic missile competition of 1959. Martin's Bold Orion (Project 199B) entrant was a 37 foot two-stage solid fuelled missile based on the Sergeant motor that was air-dropped from a B-47 jet bomber operating from Patrick AFB. Six of the twelve test flights were successful from May 1958 to October 1959, but Martin failed to win the contract when the competition was put out to tender (see Alpha Draco). Bold Orion gathered a measure of notoriety when it was fired in a simulated satellite interception against the Explorer 6 satellite on October 13, 1959 (it was a near miss). High Virgo (Project 199C) was the code name of Convair's entry into the ALBM sweepstakes. The single stage missile (again based on the XM-20 motor, as were all three entrants) was designed to be launched from Convair's flashy B-58 supersonic bomber. High Virgo (bearing the nickname "King Lofus") was flown five times at the Canaveral range from September 1958 to September 1959 but failed to bring the ALBM contract to Convair.

Bold Orion missile M-6 is shown prior to being mounted beneath the wing of a B-47 jet bomber at Patrick Air Force Base in November 1958.

Bold Orion M-6 is securely fastened beneath the right wing of the B-47 launch aircraft.

THE BLACK GOAT
The "Black Goat" was Boeing's long-nosed B-57 aerial test bed for the Bomarc interceptor program. Based at Patrick AFB in the late 1950's, the heavily modified jet bomber carried the Bomarc's guidance section in the extended nose of the aircraft, and would fly simulated intercepts on military aircraft flying in Cape Canaveral airspace. Guidance commands were transmitted to the aircraft's nose section from the Sage control center on the ground. The "Black Goat" would veer off from the target aircraft at the last possible moment, an aerial game of chicken that helped perfect the Bomarc intercept technique at the same time that actual Bomarc intercepts were conducted from Complex 4 at the Cape against QB-17 drone targets. The Goat (USAF serial number 52-01947) was a modified Martin B-57 bomber (a license-built version of the RAF's English Electric Canberra) that was later rescued from the boneyard after the Bomarc test program ended, to serve as an aerial launch platform for high altitude probe rockets in 1964.

Boeing's distinctive "Cyrano-nosed" B-57 aerial test bed for the Bomarc program, known as the "Black Goat," is shown at Patrick Air Force Base.

THE NORTH AMERICAN X-10 PILOTLESS AIRCRAFT
The North American X-10 pilotless aircraft was a technology demonstration vehicle for the Navaho cruise missile program, with the primary task of developing the automated landing system. Conventional jet engines substituted for the powerful ramjet engines of Navaho (G-26) in the 66 foot long aircraft, which was smaller than the actual Navaho missile. Flight testing of the X-10 originated at Edwards Air Force Base in California in October 1953 (15 flights) before the Air Force transferred testing to Cape Canaveral in 1955. Twenty test flights were conducted from the Skid Strip at the Cape from August 1955 to November 1956. The first autolanding at the skid strip by an X-10 aircraft occurred on February 3, 1956. Several X-10 air frames were lost in crashes, including two of the final three flights in support of the Bomarc program in 1959.

X-10 vehicle number 52-5, former Navaho test vehicle is prepared for flight in September 1958 at the Skid Strip to be utilized as a target drone for the Bomarc interceptor missile. The X-10 vehicles failed to provide a target for Bomarc and the project was cancelled after only three flights. All three X-10 drones were lost during flight or during landing operations.

HOUND DOG
North American's Hound Dog (GAM-77) cruise missile was a short range "stand off" weapon (500 miles) that was air-dropped from a B-52G bomber. Test flights of the 40 foot missile commenced on the Cape Canaveral range in April 1959, with the B-52 launch aircraft operating out of Eglin AFB on the Florida gulf coast. The Hound Dog's airframe was very similar to North American's X-10 pilotless aircraft, including canard control surfaces on the forward fuselage. The large J-52 turbojet engine was mounted in a pod carried beneath the airframe. Hound Dog was in USAF service from 1962 to 1975.

Two North American Hound Dog missiles are carried beneath the wings of a Boeing B-52G bomber.

SKYBOLT

Skybolt was a short-lived intermediate range, nuclear missile designed to be air-dropped from a B-52 bomber. The Air Launched Ballistic Missile (ALBM) was the end-result of the unusual "fly-off" competition in 1959 between the Alpha Draco, Bold Orion and High Virgo vehicles [see], collectively known as Project 199. Ironically Douglas Aircraft won the production contract for the Skybolt ALBM, but was not involved in the competition at Cape Canaveral which was contested by McDonnell, Martin and Convair. Skybolt, designated GAM-87A by the Air Force, was a 38 foot solid propellant missile with a range of 1,150 miles. The missile was test flown six times on the Cape Canaveral range from April to December 1962 before the program was cancelled by President Kennedy in December 1962. The British government was going to buy Skybolt before the cancellation, which led to the Royal Navy acquiring the Polaris A-3 missile from the Americans to comprise the UK's nuclear deterrence forces.

A Skybolt missile on its handling dolly prior to being mounted beneath the wing of the B-52 launch aircraft (background). The staging area for Skybolt tests on the Atlantic Missile Range was Eglin AFB on the Florida panhandle.

A VARIETY OF LAUNCH FACILITIES

The one distinguishing characteristic of Cape Canaveral launch facilities is (or was) their extreme variety - none of the launch pads for different rockets were alike, because every contractor arranged for different types of launch stands, umbilical masts and gantry towers for their own rockets or missiles. These facilities ranged from outright simplicity (the Army's Redstone and Jupiter pads) to extreme complexity (such as the Convair Atlas pads and gantries). Atlas' launcher mechanism, with its complicated hydraulic hold-down arms and propellant lines had no equals among the other rocket types at Canaveral. Launch mounts for most of the other rockets at Cape Canaveral simply supported the vehicle, with separate provisions for propellant lines and other umbilicals. The giant Saturns were restrained with powerful clamps that were released when engine thrust reached their maximum level. An overview of the various Cape Canaveral launch facilities is presented in the following chapter.

Simplicity: (Right) The earliest launch facilities at Cape Canaveral tended to be the simplest. Redstone and Jupiter were mounted on simple rings that could be adjusted for azimuth, with a Long Cable Mast to carry the umbilicals and coolant lines.

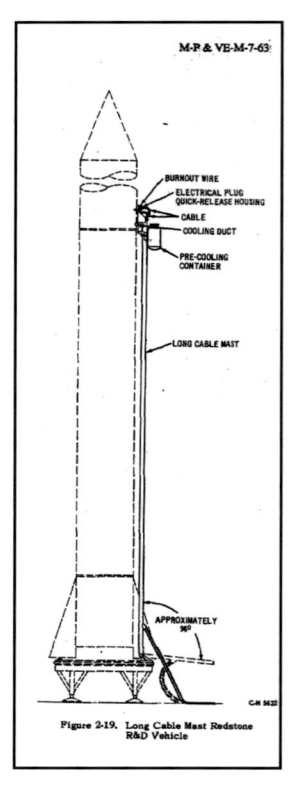

Figure 2-19. Long Cable Mast Redstone R&D Vehicle

(Left) The tactical launch mount of the Thor missile at pad 18B was identical to the operational version deployed overseas in England. A rigid mast provided umbilical connections.

(Right) The launch mount for Thor (and the early Thor-Able and Thor-Delta launch vehicles at pad 17) was a circular metal ring which supported the rocket. The ring position could be adjusted for different launch azimuths. The rings were finally replaced at Complex 17 when solid rocket motors were added to the vehicle beginning with Delta 25 in August 1964. The umbilical mast was mounted on a hinge to move the mast away from the vehicle at liftoff, pulling out the T-0 umbilicals as it did so.

(Left) An unusual view up through the opening of the mounting ring at complex 17 as a Delta first stage is lowered into position on the support posts.

(Right) The twin Thor-Delta pads at Complex 17 are illustrated in this NASA photo taken in December 1963. Delta 22 (Tiros 8) occupies pad B in the foreground, while the gantry is parked at pad A in the background. The abandoned gantry of pad 18B is visible at the far right edge.

(Below) Convair's sign off the main road points the way to the four Atlas launch complexes at Cape Canaveral in the late 1950's.

Complexity: (Right) The hold-down arms and the rest of the complicated Atlas launcher mechanism. The jaws of the hold-down system snap back at liftoff as spring-loaded doors on the rocket close to maintain the streamlined exterior of the rocket.

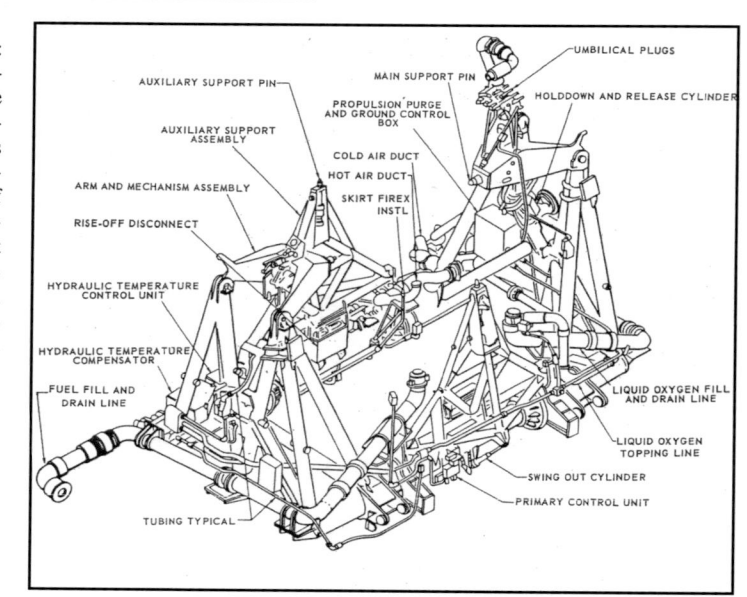

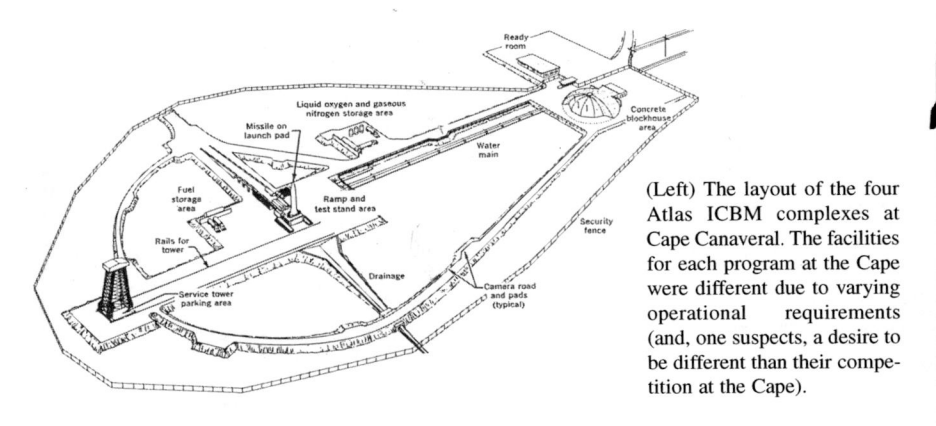

(Left) The layout of the four Atlas ICBM complexes at Cape Canaveral. The facilities for each program at the Cape were different due to varying operational requirements (and, one suspects, a desire to be different than their competition at the Cape).

There's more to the mobile service towers at the Atlas installations at Cape Canaveral than meets the eye: (Right) The Atlas service towers were parked in a position perpendicular to the launch pad rather than straight back from the pad as at most other sites.

(Left) The service tower was actually carried on a motorized Transfer Table at left that deposited the tower separately at the launch pad (background).

The transfer Table was controlled from the panel at right.

The transfer bogies at the launch pad (below) moved the tower onto the guide rails so that it could roll into position against the rocket. Complicated, but it worked well (except for the time at pad 14 when a balky diesel engine on the transfer table caused a one day delay in the launch of Gordon Cooper on MA-9).

(Right) Until the advent of the Atlas program in June 1957 Navaho's pad 9 near the Skid Strip was the largest and most impressive launch facility at Cape Canaveral. The gaping maw of the flame deflector dominates the front end of the concrete hard stand, and the erector gantry looms behind the Navaho missile before being retracted to a 45 degree parked position behind the hard stand.

(Left) In the summer of 1958 umbilical masts were added to each of the four Atlas complexes by the Air Force as Atlas A testing was concluded and the more advanced Atlas B model with two functional engines (three nozzles) was introduced for testing. Atlas A did not require an umbilical connection to the payload section as did Atlas B.

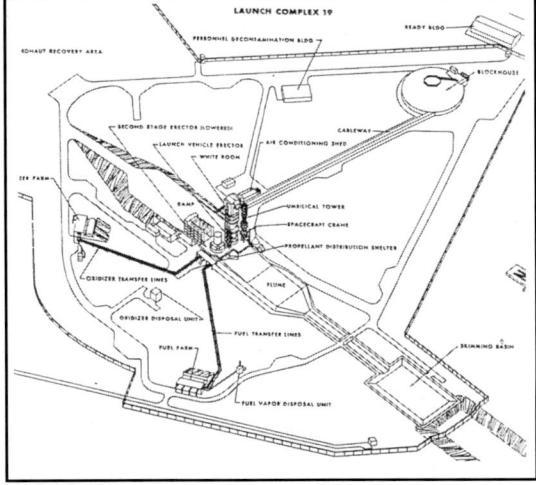

(Right) The layout of Launch Complex 19 at Cape Canaveral (modified for Project Gemini) is typical for both Titan I and II. The Titan blockhouse was 600 feet from the pad.

Note the differences in the erector gantry (left) at pad 19 compared with the Atlas layout on p.111. The umbilical tower at right is 102 feet tall.

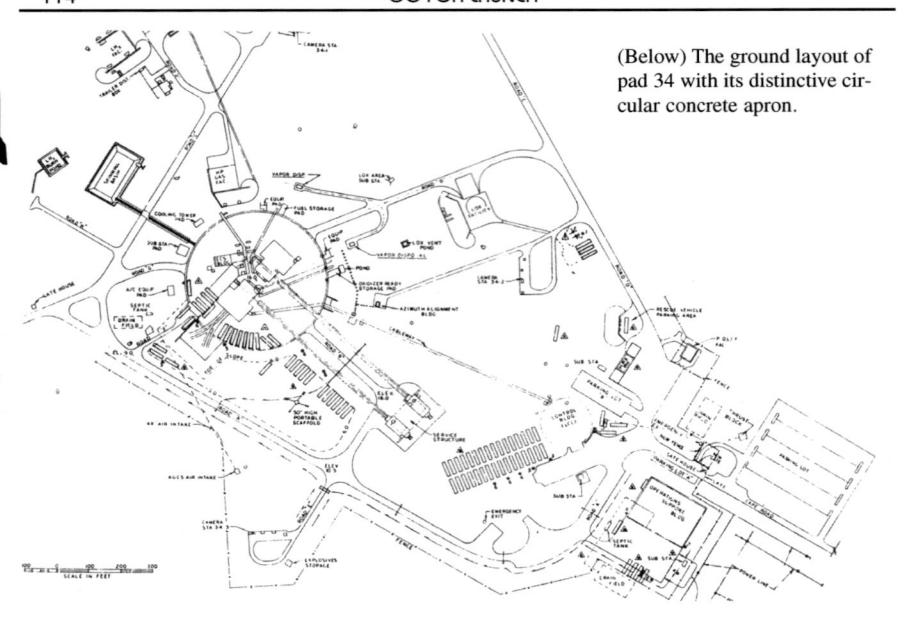

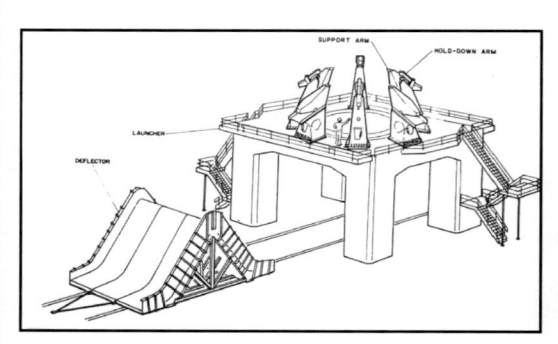

(Below) The ground layout of pad 34 with its distinctive circular concrete apron.

(Above) Everything at the Saturn complex at pad 34 seemed oversized compared to the other launch complexes at Cape Canaveral in the early 1960's. The concrete launch pedestal stood 26 feet high and was 42 feet on a side.

The relative simplicity of the Delta II pad at Complex 17A the launch platform on the umbilical tower side hangs over the flame trench which is a design feature that has been retained from the original Thor pads completed at the site in late 1956. A concrete flame duct was added to pad 17B when it was modified for Delta III operations.

The Industrial Area (missile hangars)

The Industrial Area was built up at the north end of the Canaveral base throughout the 1950's to concentrate most of the Missile Assembly Buildings (hangars) and technical support (machine shops, payload preparation) in one central location. The Range Control Building was located in the Industrial Area to oversee all launch countdowns and coordinate the downrange tracking sites. The Range Safety Officer (RSO) was also located in the Control Building (not the blockhouses), responsible for protecting Cape Canaveral workers and the surrounding communities from errant rocket launches. There has never been a civilian or military fatality or injury resulting from an aborted launch at Cape Canaveral. There are a total of 15 hangars in the Industrial Area.

Convair's missile assembly hangars J (left) and K in the Industrial Area of Cape Canaveral in the late 1950's. Behind the Atlas hangars are the Martin Titan facilities under construction (hangars T and U).

(Left) Martin technicians performing preflight checkout procedure on the first Vanguard rocket, TV-2, at Hangar S in the summer of 1957.

(Right) A Thor rocket for the ASSET program is backed into Hangar M for processing in July 1963, joining a Thor-Delta first stage already inside the building.

(Above) An Atlas rocket makes the turn into the Cape's Industrial Area after arriving by transport aircraft from San Diego at the Skid Strip. Atlas 195D was launched on February 17, 1965 for the Ranger 8 lunar mission.

(Right) The second Atlas rocket, 6A, is on the checkout fixture in Hangar K (left) in the summer of 1957.

(Left) Atlas 288D is moved by crane to the transport cradle inside the hangar, where the booster section will be joined to the main part of the missile. Mariner 4 to Mars was the payload of this rocket in November 1964.

(Right) The instrumentation section of Juno I launch vehicle number 26 (code UV) is mated to the rocket in Hangar R prior to the unsuccessful Explorer 2 launch in March 1958.

(Left) Atlas 8F is delivered to Convair's Hangar K on August 16, 1962 for launch one month later from pad 11 at Cape Canaveral.

(Right) Technicians get ready to remove a Mark IV reentry body from its handling fixture inside Hangar N, where General Electric processed Atlas and Titan nose cones.

(Left) Project Mercury workers inspect the retro-package of the MA-8 spacecraft inside NASA's Hangar S at Cape Canaveral in 1962, where all Mercury spacecraft processing took place. The Mercury astronauts also had quarters at Hangar S where they stayed during training and prior to their flights into space.

MISSION CONTROL CENTER

Before politics intervened in 1962 to locate the new Manned Spacecraft Center in Houston, Texas, NASA's Project Mercury space flights and the first manned Gemini flight were controlled from the Mission Control Center at Cape Canaveral. NASA's first Mission Director, Christopher C. Kraft and his team of mission controllers (including Gene Kranz) literally invented the mission control process from the Control Center, located in a former RCA telemetry building on Mission Control Road (where Press Site No. 1 is still located). The facility was completed in October 1960, just in time for the Mercury Redstone mission which will be forever remembered as the day they launched the escape tower (see Mercury Redstone). Engineers (controllers) occupied sixteen positions in three rows of consoles facing a big Mercury orbital map. The Mission Control Center was made famous by the public affairs commentary of Col. John "Shorty" Powers that was carried on network radio and television coverage during the first manned missions. The controller's consoles were moved to the visitor center at Kennedy Space Center in the late 1990's, and are featured in a rather cheesy display. In November 2005 NASA proposed to demolish the historic building, citing maintenance costs and safety reasons for their decision.

Mission Control Center at Cape Canaveral as it appeared in June 1964. All six manned Mercury flights were controlled from this location, as was the first manned Gemini mission (GT-3) before the new Mission Control facility was operational at the Manned Spacecraft Center in Houston.

(Below) The interior of the famous Mercury Mission Control room during Scott Carpenter's MA-7 mission in May 1962.

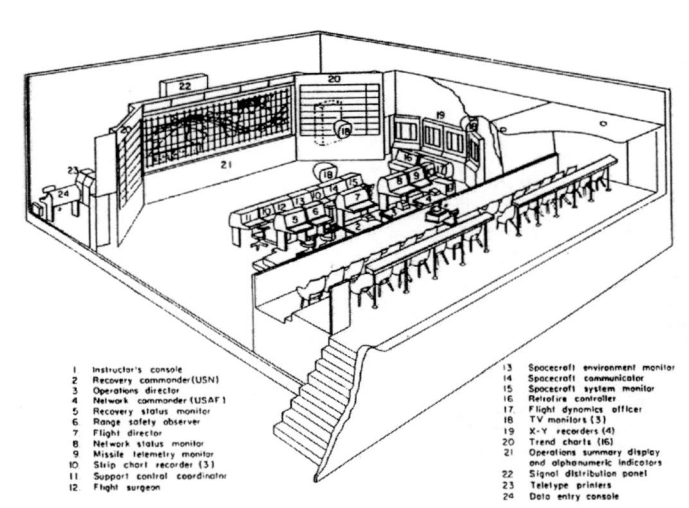

(Left) Diagram of the arrangement of consoles inside the Mercury Mission Control Center.

(Above left) Col. John "Shorty" Powers was the voice of Mission Control to millions of radio listeners and television viewers in the 1960's.

1. Instructor's console
2. Recovery commander (USN)
3. Operations director
4. Network commander (USAF)
5. Recovery status monitor
6. Range safety observer
7. Flight director
8. Network status monitor
9. Missile telemetry monitor
10. Strip chart recorder (3)
11. Support control coordinator
12. Flight surgeon
13. Spacecraft environment monitor
14. Spacecraft communicator
15. Spacecraft system monitor
16. Retrofire controller
17. Flight dynamics officer
18. TV monitors (3)
19. X-Y recorders (4)
20. Trend charts (16)
21. Operations summary display and alphanumeric indicators
22. Signal distribution panel
23. Teletype printers
24. Data entry console

Operations Room and Observation Room, Mercury Control Center (Cape Canaveral)

THE SKID STRIP
The Skid Strip runway was one of the earliest major facilities built at Cape Canaveral after it was established in 1950. It was constructed to recover Snark intercontinental range cruise missiles, which landed on metal skids rather than wheeled landing gear, hence the name Skid Strip. The runway was also used for the X-10 drone aircraft and would have hosted Navaho landings had that missile ever returned to base.

(Right) The core stage of the Titan IV vehicle that launched the Cassini probe to Saturn arrives at the Skid Strip in February 1997 aboard an Air Force C-5 transport.

(Left) Today the Skid strip serves as the airport for Cape Canaveral where most of the rockets (and their payloads) launched at the Cape begin their journeys to space.

(Below) The Skid Strip got its name from the unpiloted cruise missiles that used to be recovered there. Here a Snark missile launched on April 16, 1957 is examined on the Strip after sliding to a stop on its landing skids.

(Left) Most of the large rockets launched at Cape Canaveral have arrived by air since the late 1950's (with the exception of the Thor-Delta, which still arrives by road to this day). Here an Atlas first stage rocket is disgorged engines first from an Air Force propeller-driven C-133 transport in front of the control tower at the Strip in August 1964.

INCIDENTS AND ACCIDENTS

Cape Canaveral has had its share of spectacular failures and tragic accidents during more than 55 years of launching missiles and rockets - it is nearly unavoidable given the nature of the hazardous work undertaken there. The Air Force and NASA are very proud of their range safety record: there never has been a civilian or military casualty as a result of an errant rocket launch at the Cape. The range safety officer will not hesitate to 'terminate' a missile test or space launch if a vehicle malfunctions and heads toward an inhabited area (even if the indication is false, which has happened more than once.) Fortunately industrial accidents have been few and far between. The first fatality at the Cape occurred at an Atlas gantry on July 9, 1958 when Fred D. Adams fell from the tower to his death. Three technicians were killed on April 14, 1964 at the Delta spin test facility by an accidental ignition and firing of a solid upper stage motor, and three astronauts perished in a fire inside their Apollo 1 spacecraft during a supposedly non-hazardous test at pad 34 on January 27, 1967. Recently two construction workers died during the erection of the new Delta IV launch facility at pad 37B, the first such accidents in many years. Several of the most prominent accidents and spectacular failures at Cape Canaveral are illustrated on the following pages.

No, not fireworks: What looks like a holiday fireworks display is actually the explosion of Minuteman I missile 430 just five seconds after liftoff at Cape Canaveral on July 16, 1963. The flaming debris is reflected in the waters of Port Canaveral. Test 2080 was conducted from silo 32B, and was the last missile at Cape Canaveral to use the Azusa tracking system.

(Right) The first launch attempt of the Thor missile (test 061) ended in disaster on January 25, 1957 on pad 17B at Cape Canaveral.

(Above) The following day at pad 17B, clean up crews began removing the wreckage of Thor 101 from the blackened launch platform.

(Right) The moment of ignition for the third Navaho missile (booster 007) at pad 9 on April 25, 1957. Test 565.

(Left) The missile has fallen back onto the launch platform after one of the booster umbilicals failed to separate, triggering a cutoff signal. The vehicle rose only two feet before collapsing back onto the stand and exploding.

(Right) Navaho's propellants erupt into an enormous fireball at pad 9, one of the most spectacular accidents in the history of Cape Canaveral.

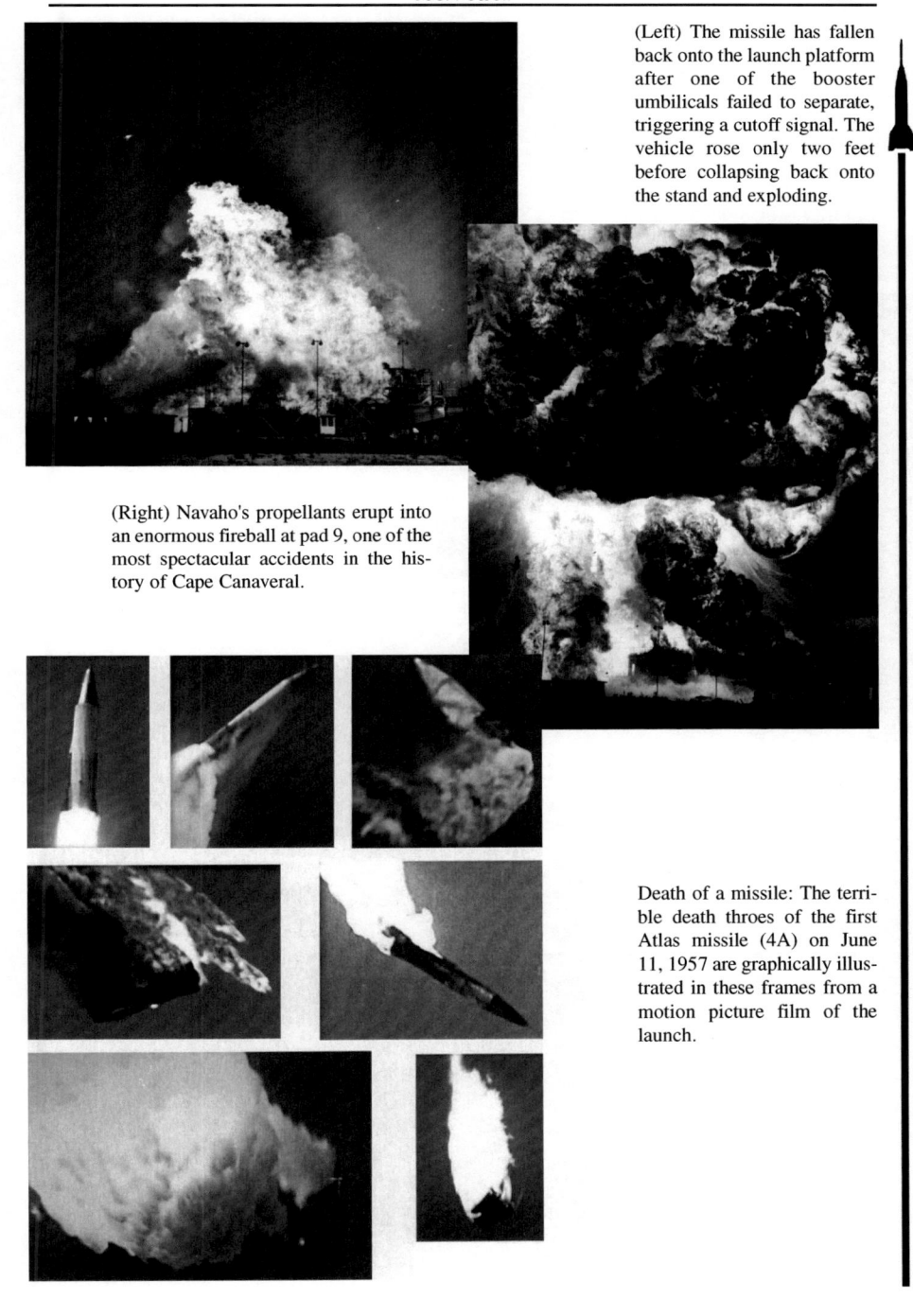

Death of a missile: The terrible death throes of the first Atlas missile (4A) on June 11, 1957 are graphically illustrated in these frames from a motion picture film of the launch.

(Right) The infamous explosion of Vanguard TV-3 at 11:44 am Eastern time on December 6, 1957. Vanguard lost thrust almost immediately after liftoff and collapsed back on the launch stand in a fiery explosion. The tiny six inch Vanguard test satellite miraculously survived the explosion, and was found in the nearby palmetto scrub, beeping ignominiously.

(Left) The remains of Vanguard TV-3 lay crumpled and burned at pad 18A. The damage would be repaired quickly enough, but not so the self esteem of the American people.

In the photo at right, Jupiter AM-23 has experienced a "structural failure" at 13 seconds into the flight on September 16, 1959 and is visibly off course. The rocket also appears to be venting something from the lower fuselage. At right the burning carcass of Jupiter AM-23 comes crashing down in the vicinity of the launch site moments later at pad 26B. There were no injuries or significant property damage in the incident.

The Atlas-Able Static Firing Disaster: (Left) The engines of Atlas 9C have just shut down prematurely after only 2.1 seconds during a static firing exercise of the first Atlas-Able lunar launch vehicle on September 24, 1959 (test 2944).

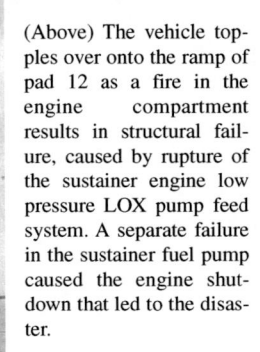

(Above) The vehicle topples over onto the ramp of pad 12 as a fire in the engine compartment results in structural failure, caused by rupture of the sustainer engine low pressure LOX pump feed system. A separate failure in the sustainer fuel pump caused the engine shutdown that led to the disaster.

(Left) Smoke pours from the stricken launch stand at Complex 12 after fires were extinguished.

(Right) The fire and subsequent explosion wrecked Complex 12. It took contractors 68 days to repair the damage.

Aftermath of the disaster: (Above left and right) These Convair photos show the devastating aftereffects of the Atlas 9C accident at pad 12. Burned and crumpled remains of the Able stage were thrown free of the main body of the rocket by the explosion.

(Left) Shattered remains of the Atlas sustainer engine lay in a heap at the foot of the launch pad, as Convair engineers look on. Convair and NASA had to transfer to pad 14 to attempt the second Atlas-Able mission in November 1959.

This Juno II orbital launch attempt in July 1959 (test number 2000) is one of the most memorable accidents in the history of Cape Canaveral. The vehicle plunged out of control and was destroyed only 5 seconds after liftoff from pad 5. Badly charred remains of the largest cluster of solid motors is shown above.

(Right) The wayward flight of Atlas 3D on April 14, 1959. The LOX and fuel fill-and-drain valves failed to release properly at liftoff from pad 13, and the damaged Atlas was translating sideways as it took off, before the range safety officer destroyed the first Atlas D model when it reached a safe altitude for termination.

(Right & Top opposite) With Launch Complex 12 already out of commission due to the Atlas-Able static firing disaster in September 1959, Convair and the Air Force came close to a crisis in the opening months of 1960 as two more launch stands were disabled by explosions less than a month apart. Had the accidents occurred even a year earlier, the disruption to Atlas testing could have seriously delayed the achievement of operational readiness at Vandenberg AFB in September 1959. These photographs depict the demise of Atlas 51D on March 10, 1960 at pad 13 (test 775). The missile was a victim of "unidentified combustion instability" at liftoff that produced loss of thrust and explosion of the vehicle.

(Left) Two consecutive Titan I vehicles exploded on the pad in 1959 during the first flights of both the lot B and lot C missile series. Missile B-5 was destroyed on pad 19 when vibration caused the hold-down bolts to detonate prematurely on August 14, 1959. The engines shut down and the vehicle was destroyed on the pad. The photo depicts the conflagration when missile C-3 was lost (test 3528) with inadvertent triggering of the destruct system before liftoff. The Titan I test program regained its momentum in 1960 with a total of 20 launches from the four Titan I test stands.

(Right) A mangled heap of oily engine turbo machinery and a crumpled engine nozzle litter the grounds of Complex 13 in the aftermath of the 51 D explosion.

(Left) On April 7, 1960 it happened again. Combustion instability in the booster engine brought down Atlas 48D before it could even achieve liftoff from pad 11, earning a NO TEST rating in the Convair record books. Launch Complex 12 was returned to service on May 20, 1960 with the launch of Atlas 56D, alleviating pressure on the test schedule, leaving only two pads out of three not in service. Complex 14 was reserved for space launches including the Mercury manned flights.

(Right) The blackened remains of Atlas 48D's booster section lies askew in the pad 11 launcher mechanism on the morning after the conflagration. With structural integrity intact, engineers were able to restore pad 11 to service in only two months - Atlas 54D was launched there on June 11, 1960.

Clips from a motion picture sequence of the Mercury-Atlas 3 abort on April 25, 1961. The escape system is triggered (upper right) and the capsule descends by drogue chute (lower left) as the discarded escape tower (three dots) departs the scene. The MA-3 capsule descends under the main chute (lower right) and was recovered undamaged on the beach at Cape Canaveral was re-used for the following launch (MA-4).

(Right) After more than a year in residence at pad 36A undergoing a seemingly endless series of preflight tests, the first Atlas-Centaur vehicle (F-1) finally made it into the sky on May 8, 1962 (test 5461).

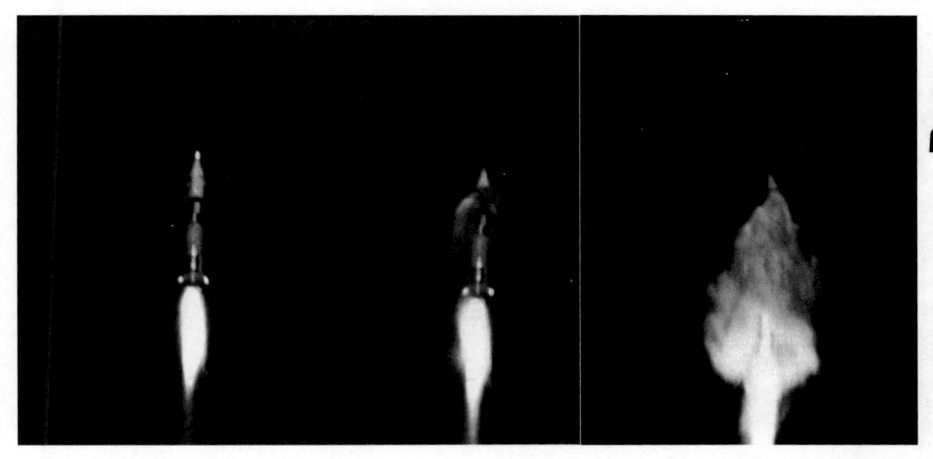

Above left, it appears that the weather shield on the Centaur stage has broken away in the slipstream. In the center picture the Centaur propellant tank has clearly ruptured and is venting liquid. At right the vehicle is engulfed in flames as the propellants are ignited by the rocket exhaust at 54 seconds after liftoff, destroying the vehicle in a manner reminiscent of the shuttle Challenger accident nearly twenty four years later. As a result of engineering changes incorporated in the Centaur design, it was four long years before NASA declared Atlas-Centaur operational.

These two photographs reveal the interior of the Spin Balance Facility at Cape Kennedy's Area 39 after a devastating fire took the lives of three technicians on April 14, 1964. The X-248 solid motor upper stage for the Orbiting Solar Observatory - B satellite was accidentally ignited by a stray static discharge, ricocheting off the ceiling and flying down to the other end of the spin test facility. The ruined OSO payload is visible in the photo above (with the 'caution' sign). Safety procedures were tightened as a result of the accident, and there has never been another accidental motor ignition at Cape Canaveral.

(Below) Atlas-Centaur 5 (AC-5) has fallen back onto the launch stand and is already in flames as the first stage cut off prematurely only two seconds after liftoff on March 5, 1965.

(Above) Like a bomb going off at pad 36A, the entire load of AC-5's propellants detonated at 8:35 am EST. Note the two dimpled metal panels that were forced inwards as air was sucked from the area when the fireball erupted (vacuum implosion).

(Lower left) Even the paint on the umbilical tower was afire as the rocket propellants burned after the explosion.

(Below) The shattered nose fairing of AC-5 lies at the foot of the ruined Centaur launch complex.

(Right) Gemini 6A's first attempt to join Gemini 7 in orbit for critical rendezvous exercises ended in a hair-raising launch abort at Cape Kennedy on December 12, 1965. Ignition occurred on schedule at 9:54 am, but 1.2 seconds later the engines were shut down by the malfunction detection system when an umbilical plug ejected prematurely. Closer examination of engine telemetry records by Aerojet engineers indicated that the thrust began to decay before the plug dropped out, and an inspection of the engines revealed the infamous dust cover that blocked a thrust inlet. The dust cover was removed and GT-6A was launched successfully three days later to complete the historic orbital rendezvous with Gemini 7.

(Right) A picture of dejection, Gemini 6A astronauts Tom Stafford (left) and Wally Schirra walk down the ramp at pad 19 after the abort.

(Left) Astronauts Gus Grissom (left), Ed White (center) and Roger Chaffee were killed during a "routine" test inside their Apollo 1 command module at Complex 34 on January 27, 1967.

(Right) On January 27, 1967. Part of the Boost Protective Cover was removed from the spacecraft before the test to allow electrical umbilicals to be attached. An electrical short circuit, believed to have originated in exposed wires beneath the couch of Gus Grissom, ignited a fire in the pure oxygen atmosphere inside the sealed spacecraft shortly before the end of the test at 6:39 pm EST. The fire devastated the Cape Canaveral space community, and the after effects still linger today with residents and former space workers who have never forgotten the tragedy.

(Below) The fire-blackened command module number 012 is gently lowered to a flatbed truck at pad 34 on February 17, 1967 for transport to the Pyrotechnic Installation Building for examination by the Apollo 204 Review Board.

(Right) A week later the spacecraft 012 service module and adapter were demated from the Saturn IB at pad 34 and taken to the Manned Spacecraft Operations Building at KSC for post-accident analysis.

(Left) Solid rocket motor number 1 lies at an awkward angle on the deck of the launch platform at pad 17A after the motor broke loose from the Delta model 3914 core vehicle on May 18, 1977. The half-inch attachment bolt had failed, dropping the 23,900 pound solid motor into the flame bucket and tearing a 3-4 foot gash into the side of the Delta's oxygen tank. The motor was scrapped and the Delta was sent back to the McDonnell Douglas factory to replace the oxygen tank. Launch of the European Space Agency's Orbital Test Satellite, set for June 16, was rescheduled for September 1977 on another Delta booster.

(Above and Left) The Orbital Test Satellite (OTS) never made it into orbit. Just 54 seconds after liftoff on September 13, 1977, one of the Castor-4 solid rocket motors on Delta 134 ruptured and exploded almost instantly, scattering the other eight solid motors into the sky. Next up on the NASA launch schedule was an Atlas-Centaur at the end of the month, and in an eerie flashback to the early days of rocketry at the Cape, the same fate befell AC-43 (see the following page).

A promising beginning, but... (right) Atlas-Centaur 43 thunders off pad 36A in what looks like a fine start to the mission on the evening of September 29, 1977. Test 2050 was soon in trouble as a fire in the tail of the Atlas caused the vehicle to tumble wildly out of control (below left). The Centaur stage separated prematurely and the payload fairing shattered (bottom left). 55 seconds after liftoff the rocket erupted in a tremendous fireball, the second such failure in two weeks at the Cape (see Delta 134, previous).

Tridents behaving badly: (right) The 17th land pad launch of the Trident I missile at pad 25C went awry immediately after liftoff on December 16, 1978. The range safety officer had to destroy the vehicle after only 5 seconds because a procedural error resulted in a liftoff with the guidance system shut down.

(Above) The first submerged launch of a Trident II (D-5) missile went terribly wrong as the missile emerged from the sea and spiraled out of control after solid motor ignition. The vehicle self-destructed after only 4 seconds on March 21, 1989 (test 1934) after the nozzle was damaged by water overpressure during ejection from the submarine. Strengthening the nozzle was successful in preventing further occurrences of this problem.

(Below Right) The punctured Centaur hydrogen tank of AC-68 is lowered from the gantry at pad 36B after a work platform perforated the thin-skinned tank on July 13, 1987. The extent of the damage to the pressure-supported stainless steel skin of the Centaur is evident below. It would be more than two years before AC-68 would take to the skies because there were no spare tanks to replace the damaged upper stage, which was the last of the NASA-sponsored Centaur missions before commercial Atlas launch operations got under way in 1990. AC-68 was finally launched on September 25, 1989.

Spectators at Cape Canaveral were stunned when Delta 241 erupted in a tremendous explosion just 13 seconds after liftoff on January 17, 1997 at pad 17A. A catastrophic failure in strap-on solid motor number 2 destroyed the Navstar satellite payload and the rocket at 1,589 feet, one of the closest accidents to the ground at Cape Canaveral since the 1970's. Dutch space watcher Bart Hendrickx took these pictures from Press Site 1 which was uncomfortably close to the fireball (a TV camera at the launch site showed burning solid propellant fragments raining down from the sky for several minutes after the blast) There were no injuries, but property damage was extensive at the launch site (parked vehicles) and at the adjacent Missile and Space Museum.

DOWNRANGE OPERATIONS

"America's 6,000 Mile Shooting Gallery." This is how National Geographic magazine described Cape Canaveral's far-reaching network of twelve tracking stations (see opposite) extending downrange to Ascension Island in the South Atlantic. Station 1 was Cape Canaveral itself. The installations were all linked to the Cape by undersea cables to transmit the tremendous amounts of data generated during each missile test. The magnetic tapes with the data were also rushed by air back to Canaveral to provide the original recordings for the analysts back at the Cape. Construction on the downrange stations was undertaken by the Air Force beginning in 1951 on Grand Bahama Island, followed by Eleuthera, San Salvador, Mayaguana and Grand Turk. In 1956, additional stations were activated downrange as the missiles (Snark in particular) flew increasingly long distances. Finally, as the range for Atlas and Titan ICBM's improved to at least 9,000 miles, an additional station was added at Pretoria in South Africa. Extra telemetry coverage was provided by eleven tracking (picket) ships. During the 1960's several of the stations were closed, reducing the number to nine, and then the range was largely dismantled in the 1970's as missile testing at Cape Canaveral was mostly concluded, except for Navy programs like Poseidon and Trident.

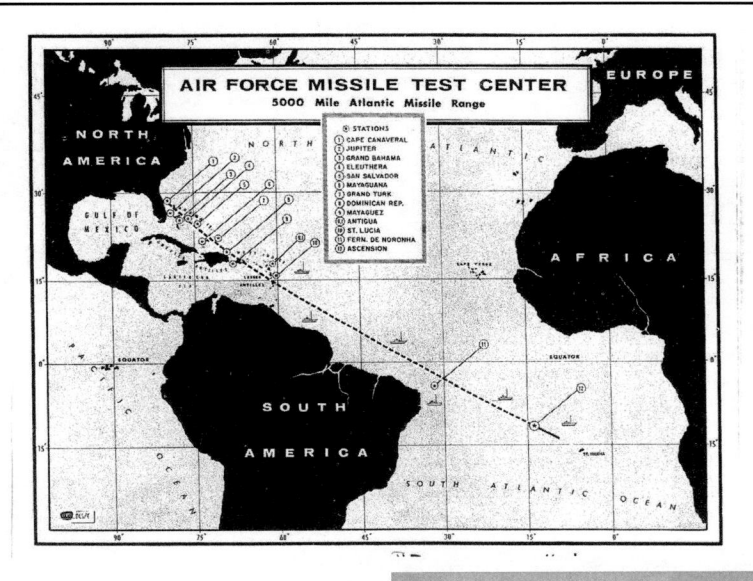

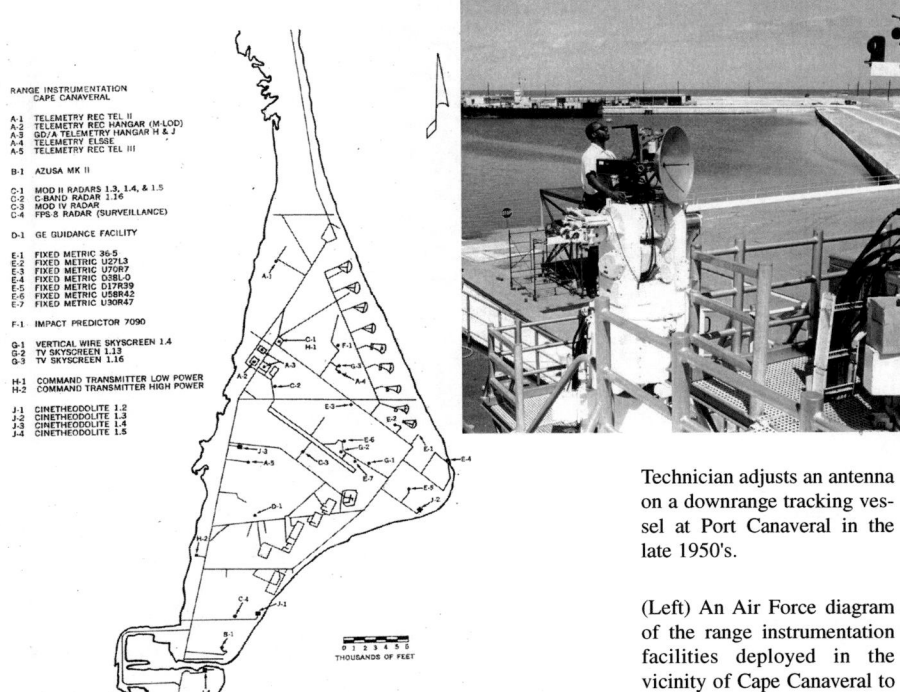

AIR FORCE MISSILE TEST CENTER
5000 Mile Atlantic Missile Range

STATIONS
1. CAPE CANAVERAL
2. JUPITER
3. GRAND BAHAMA
4. ELEUTHERA
5. SAN SALVADOR
6. MAYAGUANA
7. GRAND TURK
8. DOMINICAN REP.
9. MAYAGUEZ
10. ANTIGUA
11. ST. LUCIA
12. FERN. DE NORONHA
13. ASCENSION

RANGE INSTRUMENTATION
CAPE CANAVERAL

A-1 TELEMETRY REC TEL II
A-2 TELEMETRY REC HANGAR (M-LOD)
A-3 GD/A TELEMETRY HANGAR H & J
A-4 TELEMETRY ELSSE
A-5 TELEMETRY REC TEL III

B-1 AZUSA MK II

C-1 MOD II RADARS 1.3, 1.4, & 1.5
C-2 C-BAND RADAR 1.16
C-3 MOD IV RADAR
C-4 FPS-8 RADAR (SURVEILLANCE)

D-1 GE GUIDANCE FACILITY

E-1 FIXED METRIC 36-5
E-2 FIXED METRIC U27L3
E-3 FIXED METRIC U70R7
E-4 FIXED METRIC D3BL-O
E-5 FIXED METRIC D17R39
E-6 FIXED METRIC U58R42
E-7 FIXED METRIC U30R47

F-1 IMPACT PREDICTOR 7090

G-1 VERTICAL WIRE SKYSCREEN 1.4
G-2 TV SKYSCREEN 1.13
G-3 TV SKYSCREEN 1.16

H-1 COMMAND TRANSMITTER LOW POWER
H-2 COMMAND TRANSMITTER HIGH POWER

J-1 CINETHEODOLITE 1.2
J-2 CINETHEODOLITE 1.3
J-3 CINETHEODOLITE 1.4
J-4 CINETHEODOLITE 1.5

0 1 2 3 4 5 6
THOUSANDS OF FEET

Technician adjusts an antenna on a downrange tracking vessel at Port Canaveral in the late 1950's.

(Left) An Air Force diagram of the range instrumentation facilities deployed in the vicinity of Cape Canaveral to photograph and track missile launches.

(Right) One of the two big ANQ-18 telemetry receiver antennas at Cape Canaveral, Station No. 1 of the Atlantic Missile Range in the 1950's and 1960's. (Below Left) The antenna field of the Azusa Mark II range safety tracking system was located on the south end of the base near the main gate. (Below right) The distinctive Azusa antennas were mounted on external "tripods" on each Atlas missile (and some Atlases even carried two of them).

Downrange recovery forces: When recovery was required of a payload launched on a missile from Cape Canaveral, the Air Force would deploy recovery vessels downrange. (Right) A sailor brings aboard a data capsule ejected from the reentry vehicle of Titan I J-8 on September 28, 1960.

(Left) A whaleboat dispatched from the recovery ship attaches a line to the RVX-2A reentry vehicle floating nearby in October 1960.

Scenes from the recovery of Gemini 9A and 11 in 1966. A Navy task force was required to retrieve the astronauts who landed in the West Atlantic Recovery Area 46-1. The recovery carrier (visible on the horizon) was the USS Wasp.

THE RECOVERY OF LIBERTY BELL 7

In a remarkable story of perseverance against the odds, undersea explorer Curt Newport located and then recovered the sunken Mercury spacecraft, Liberty Bell 7, in 1999 after a search effort lasting fourteen years. Newport returned the barnacle encrusted capsule from the Atlantic Ocean to Port Canaveral (below) on July 21, 1999. It was the 38th anniversary of Gus Grissom's launch on Mercury Redstone 4 at nearby pad 5 from Cape Canaveral.

(Above) Pilot Jim Lewis in Marine helicopter 32 (call sign: 'Hunt Club 1') struggles to lift Liberty Bell 7 as it filled with water after the hatch inadvertently blew off shortly after landing. Astronaut Gus Grissom was eventually rescued, but his capsule sank to a depth of 15,600 feet, where it stayed until Curt Newport rediscovered it in 1999.

(Right) Reporters examine the newly recovered Liberty Bell 7 capsule in Port Canaveral on July 21, 1999.

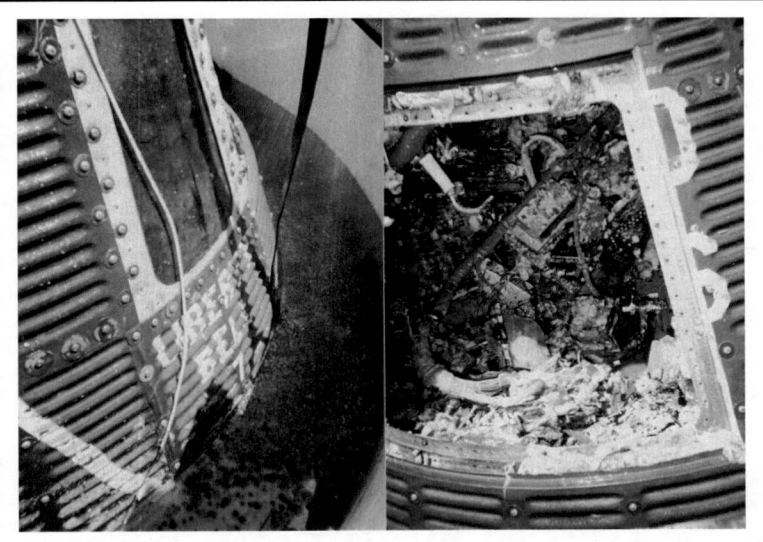

(Upper left) The letters of Liberty Bell 7 (and a "crack" to the left) were still intact after thirty eight years on the bottom of the Atlantic. (Upper right) A layer of sludge fills the bottom of the capsule as seen through the infamous open hatch of Liberty Bell 7. Many of the plastic components in the cabin survived immersion perfectly intact, while most of the metal parts were badly corroded.

(Right) The fully restored capsule on display at Kennedy Space Center in 2000, with the Mercury pad leader Guenter Wendt (second from left), Marine helicopter pilot Jim Lewis, and Gus Grissom's older brother Lowell (at right).

AIR FORCE SALVAGE CREWS
Prior to every missile launch from Cape Canaveral in the 1950's and 1960's, Air Force crash boats would be deployed from Port Canaveral to wait off shore in case of mishap. Divers aboard the boats would head for the bottom to try and retrieve missile parts (usually from the engine) to assist the accident investigators. Often salvaged components were the only evidence of what went wrong on a failed flight. The crash boats would be guided to the impact sites off shore by theodolite and radar measurements from the Cape, and sometimes by observers stationed at the top of the famous lighthouse. Lou Berger Divers Inc. was contracted by the Air Force in 1954 to conduct the salvage missions in the early days of Cape testing. The salvage crews had a particularly apt nickname: "The Undertakers."

(Right) Debris from the B-1 booster engine of the second Atlas flight test (6A) is hauled aboard a salvage boat just off the beach of Cape Canaveral within sight of the launch pad. The recovery was performed on October 7, 1957 about two weeks after the errant Atlas launch.

(Below) Engine debris from Atlas 6A is lifted ashore at Port Canaveral as salvage crews worked to discover the cause of the second Atlas failure.

(Right) NASA accident investigators pieced together the remains of the Mercury capsule destroyed in the failure of MA-1 on July 29, 1960.

(Left) Sailors aboard the destroyer USS Dupont made a surprise discovery downrange from Cape Kennedy on August 19, 1965 when they recovered a 17 foot section of the Titan II launch vehicle of Gemini 5, the first and only time a portion of the launch vehicle was retrieved after a manned launch on a ballistic rocket. The Titan segment is currently on display at the Alabama Space and Rocket Center in Huntsville where it has resided for many years.

PHOTOGRAPHIC SERVICES AT CAPE CANAVERAL

When the Air Force hired Pan American Airways to provide security services at Cape Canaveral in 1954, RCA Corp. was also signed on to provide "technical" support for the range, which included photographic services to document the launches and related activities, and operating the base's photo processing laboratory. Later Technicolor was brought on board to produce motion picture records of prelaunch testing. RCA technicians were responsible for setting up the high speed motion picture and sequential still cameras that recorded each launch for the technical analysts at Patrick AFB. RCA also tended to the photo-theodolites and telescopic tracking cameras that recorded powered flight and staging of each rocket. This included the three IGOR tracking cameras and the 500 inch focal-length ROTI telescope, along with the powerful Sandia tracking camera mounts. RCA also had a large staff of still photographers that recorded all aspects of work at Cape Canaveral that are featured in this book. These photographs were provided to the news media by the Air Force or NASA Public Affairs Offices. One of the RCA photographers was Chuck Rogers, who was well known for his artistic approach to missile photography, often featuring the interplay of floodlights and liquid oxygen vapor at night. He was also renowned for his time-lapse night time portraits of service tower rollbacks. RCA's photographers used large format cameras (70 mm and 4 X 5 inches). After many years of maintaining separate photographic contractors, the Air Force and NASA have recently amalgamated their photographic contractors at Kennedy Space Center and Cape Canaveral. NASA will now provide photographic services for the entire range through its contractor, Johnson Controls.

An RCA camera operator on his CZR camera platform follows the progress of a Titan II rocket shortly after liftoff from pad 15 on February 26, 1964.

(Right) An RCA technician adjusts the cable for one of the high-speed motion picture cameras aimed at the Atlas A missile in the background in early 1958.

(Left) A series of high speed motion picture cameras are set up on permanent mounts at the foot of an Atlas launch stand at Cape Canaveral.

(Right) RCA technician sets up remote still cameras on the ramp at pad 13. Atlas 4E was launched there on November 29, 1960.

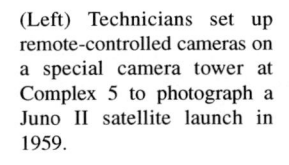

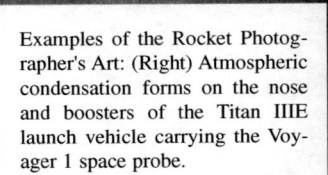

(Left) Technicians set up remote-controlled cameras on a special camera tower at Complex 5 to photograph a Juno II satellite launch in 1959.

Examples of the Rocket Photographer's Art: (Right) Atmospheric condensation forms on the nose and boosters of the Titan IIIE launch vehicle carrying the Voyager 1 space probe.

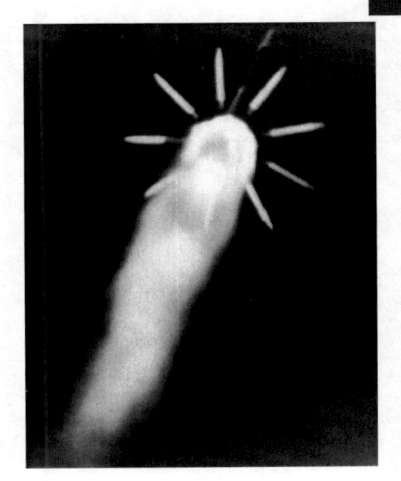

(Left) The flower petal effect is actually all nine solid motors being jettisoned from Delta 116 on October 16, 1975 during the GOES-1 weather satellite launch.

(Right) A time-exposed "streak" shot of Mariner 6's ascent, photographed from the town of Cocoa Beach on February 24 1969. The streak shot from Cocoa Beach was a favorite subject among RCA photographers working at the Cape.

148 GO FOR LAUNCH

Part II

THE SPACE LAUNCH ERA (1958-1986)

SIGNIFICANT PAYLOADS

Many historic spacecraft have begun their voyages of discovery from the launch pads of Cape Canaveral, from the early Pioneer moon shots to Ranger, Mariner, Surveyor, Viking, Voyager and Cassini. Every planet in the solar system has been a target for NASA's remarkable exploration robots, from sun-baked Mercury to distant, frozen Pluto, and they were all checked out, tested and launched from Cape Canaveral (or occasionally from neighboring Kennedy Space Center). Numerous other scientific and applications payloads have also been orbited from Cape Canaveral, including the first communications and weather satellites, and the remarkable astronomy missions such as WMAP and Swift. In the following pages many of these robotic space explorers are depicted in the various check-out and test facilities of Cape Canaveral.

(Left) In vivid contrast to the clean rooms and protective clothing of payload processing for today's space missions, technicians prepare the first Pioneer lunar probe in the open air at the top of pad 17A's gantry in August 1958. Technicians are installing the small solid fuel vernier rockets at the base of the Pioneer spacecraft.

(Right) JPL technicians perform prelaunch checks on the early Block II lunar probe, Ranger 3 (the white sphere was a balsa wood hard-landing capsule) inside Hangar AE at Cape Canaveral in December 1961. Both these payloads were unsuccessful.

(Left) Workers slowly guide a Mark 3 re-entry vehicle into position on the top of an Atlas D missile in this photo dated June 3, 1960.

(Right) Nattily attired NASA engineer handles the solar panel of the first Interplanetary Monitoring Platform (Explorer 18) in the white room of the Pad 17B gantry in November 1963. Note the slender Thor-Delta payload fairing in the background.

(Left) Surveyor spacecraft is about to be encapsulated within the Centaur payload fairing inside Hangar AO at Cape Canaveral prior to mating with the launch vehicle at Complex 36.

(Right) Mating of the Titan Centaur 12 foot large diameter payload fairing with the Viking 1 Mars probe in July 1975 inside the Spacecraft Assembly and Encapsulation Facility (SAFE-1) at Kennedy Space Center. The two Viking spacecraft achieved the first unmanned landings on Mars in the summer of 1976.

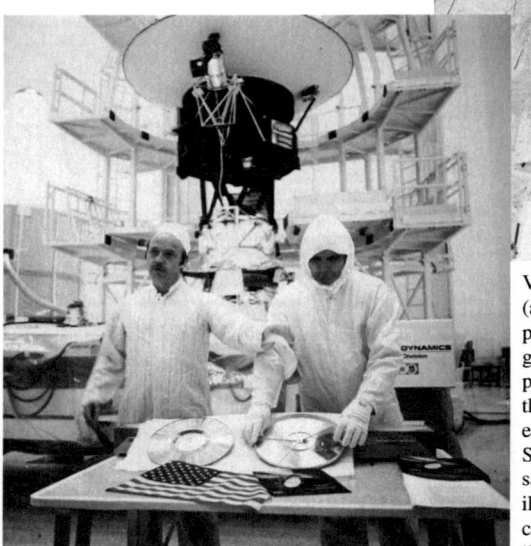

Voyager spacecraft manager John Cassani (at left) and a spacecraft technician prepare to pack the "Sounds of Earth" phonograph record and an American flag inside a protective cover prior to placing it aboard the Voyager 2 space probe. The record and engraved cover plate, developed by Carl Sagan, were intended to serve as a "message in a bottle" to any extraterrestrial civilization that might encounter the spacecraft as it drifts through our galaxy for millions of years in the future.

(Right) Technicians begin to fold the landing petals of the Mars Pathfinder spacecraft into flight position. The Sojourner rover is attached to the right-hand petal. The deflated airbags that would cushion the lander upon its arrival on the Martian surface July 4, 1996 are visible below the landing petals.

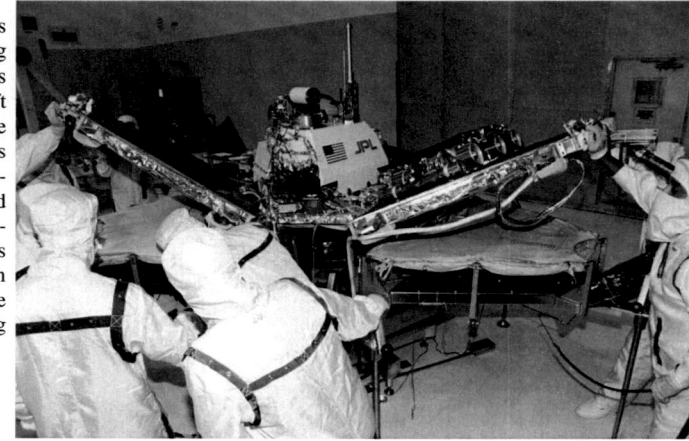

(Left) The Cassini space probe (shown in the Payload Hazardous Servicing Facility at Kennedy Space Center) is currently orbiting Saturn to explore the mysterious rings and satellites of the sixth planet from the Sun. Charley Kolhase (left), Cassini science and mission design manager, and Richard Spehalski, JPL's Cassini program manager, hold a special DVD with the scanned signatures of 616,400 earthlings. The symbolic payload was tucked behind a piece of insulation blanket on the spacecraft adorned with a Saturn and quill pen symbol (visible behind Kolhase's head). Cassini was launched on a Titan IVB rocket from pad 40 at Cape Canaveral on October 15, 1997.

(Right) The Mars Rover Spirit is inside the protective aeroshell atop the Delta II launch vehicle. The spacecraft is being enclosed within the payload fairing in May 2003 inside the white room of the pad 17A gantry.

JUNO I

A modified Jupiter-C vehicle became the basis of the U.S. Army's Juno I satellite launcher in 1957 with the addition of a fourth stage solid motor. The Redstone first stage propellant mix was also modified, a development (Hydyne) that gave rise to the enduring myth of "secret rocket fuel." The new vehicle configuration produced 83,000 pounds of thrust and stood 68.5 feet tall. After authorization from President Eisenhower in November 1957, the first Juno I was shipped to Cape Canaveral for a satellite launch attempt in January 1958. A successful launch on January 31, 1958 put America's first satellite, Explorer 1 into orbit. Despite the apparent simplicity of the Juno I system, it achieved only 3 successes during the six orbital attempts between January 31 and October 22, 1958. Juno I was retired after failures of the last two vehicles. The same upper stage problems would plague the follow-on Juno II in 1959-1961. Juno I's defining characteristic was the spinning upper stage tub (see Jupiter-C) that was launched without a protective shroud. Except for certain manned spacecraft, American launch vehicles never flew without payload fairings again.

The historic moment of liftoff for America's first satellite at 10:48 pm EST on the evening of January 31, 1958. The location was Complex 26A at Cape Canaveral. The code letters UE signify Redstone missile 29.

(Left) The first Juno I launch vehicle is surrounded by the Army gantry at Launch Complex 26A prior to the first American satellite launch in January 1958.

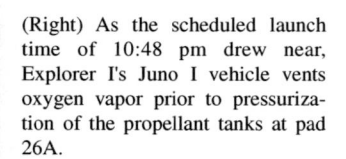

(Right) As the scheduled launch time of 10:48 pm drew near, Explorer I's Juno I vehicle vents oxygen vapor prior to pressurization of the propellant tanks at pad 26A.

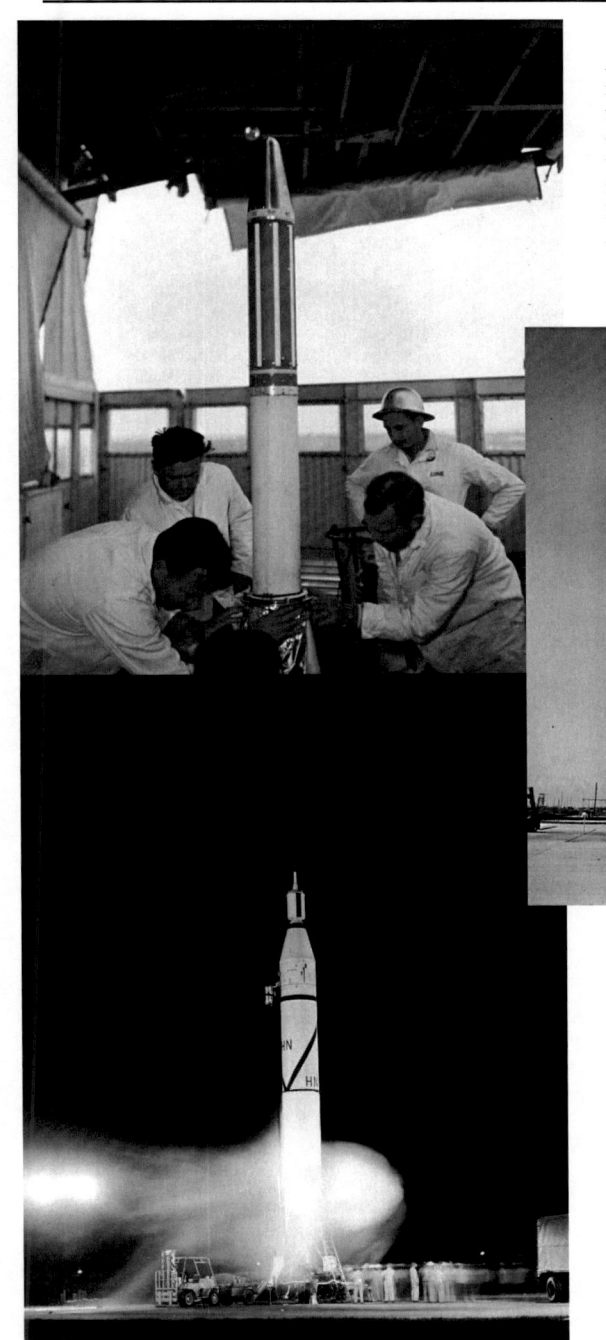

JPL technicians make final adjustments to the Explorer I payload in the open-air 'green room' at the top of the pad 26A gantry. History would be made that evening as the first American Earth satellite was launched into orbit on the Army's Juno I rocket on January 31, 1958.

(Above) Juno I TT (Redstone booster number 44) is nearly ready to launch on July 26, 1958 into orbit with Explorer IV. The was the third and last success for the ABMA in the 6 vehicle program. LC-5 supported the flight for the 38 pound radia-tion study satellite.

(Left) The last of the radiation study satellites is readied for launch from LC-5 on August 24, 1958. This payload weighed in at 38 pounds, up from the original 31 pound Explorer I payload.

Launch of Explorer V on August 24, 1958 was off to a good start but after shutdown of the Redstone first stage the separated upper stage assembly was bumped by the Redstone due to residual thrust. The mis-oriented section fell into the ocean. Two months later on October 22 the last Juno I failed in flight when the spinning upper stages broke away during ascent.

JUNO II

Juno II was only the third American space launch vehicle to be fielded after Juno I and Vanguard. Based on the Jupiter intermediate range missile, Juno II was developed by the U.S. Army to boost two early space probes to the moon before being used to launch earth satellites for the new National Aeronautics and Space Administration (NASA) from 1959 to 1961. Three solid fuelled upper stages comprised the "velocity package" for Juno II, with a total of 15 solid motors. The Juno II was an unsuccessful launch vehicle, with only four successes out of ten launches. Juno II flew from both Complex 26 and 5/6.

A Juno II launch vehicle (AM-19B for Beacon Explorer) is prepared for launch on pad 26B at the Cape in August 1959.

A Juno II launch vehicle is prepared for launch inside Hangar D. The first stage undergoes checkout by Army technicians (left), while the fairing for the upper stages is stored temporarily in the aisle adjacent to the first stage (below).

(Below left) The velocity package of Juno II launch vehicle, below right AM-19C (code PEC) is lifted to the top of the gantry at LC-26 in preparation for mating to the launch vehicle below right.

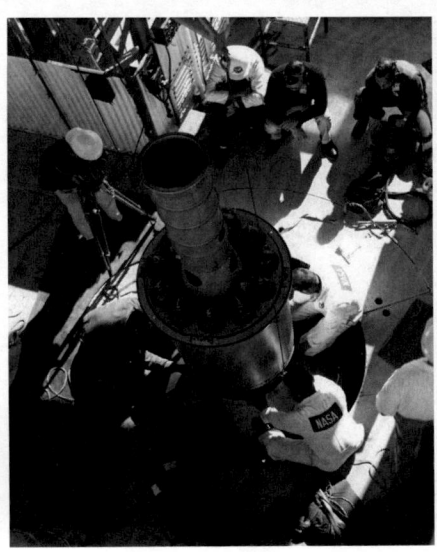

(Right) The fairing is hoisted to the top of the gantry to enclose the Juno II velocity package (upper stages) and payload.

(Below) Swirling LOX vapors add a surrealistic look to the Juno II propellant loading process at the launch pad.

History in the making: the first Juno II vehicle (left) is prepared for the U.S. Army's first lunar probe in December 1958 on behalf of the newly constitute space agency (NASA). Liftoff appeared flawless (right) but a velocity shortfall from the upper stages saw the tiny Pioneer probe climb to only 63,000 miles before falling back to Earth.

VANGUARD

The first American satellite program was announced in July 1955 and was assigned to the Naval Research Laboratory as Project Vanguard. Two Viking rockets and 12 Vanguard launch vehicles were flown between December 1956 and September 1959 from Launch Complex 18A. Of the eleven attempted satellite launches, 3 were test versions and 8 were the full-size scientific satellites as intended for the American participation in the International Geophysical Year (1957- 1958). Early Vanguards were designated as Test Vehicles (TV-0 through TV-5) while the last seven flights constituted the Satellite Launch Vehicle (SLV) version. Only three of the 11 orbital attempts were successful. Vanguard was a highly efficient design, but it lacked adequate funding or the political commitment that was extended by the government to military space projects. The U.S. Navy (through the NRL) was relieved of its responsibility for Vanguard when NASA opened for business on October 1, 1958 and absorbed the project into the civilian space effort. NASA flew the final four Vanguards in 1959 after a four month hiatus to remedy several technical problems.

Vanguard TV-3 shown during a propellant loading test at launch complex 18A in late 1957 prior to the first attempted United States satellite shot on December 6, 1957.

Viking number 13 (one of two remaining sounding rockets flown by the NRL in the 1949-1955 era at White Sands, New Mexico) is shown at the incomplete pad in 1956 just before its flight in December to prove out the facilities and range for Vanguard. This rocket was designated as Vanguard Test Vehicle Zero or just TV-0. The launch was completely successful.

TV-0 is shown beneath a cloudy evening sky at the Cape, enclosed by the transplanted service tower which had been salvaged from the Viking launch site at White Sands. This was a major cost-savings measure for the austerely funded Vanguard program. The second Viking (#14 or TV-1) was flown from LC-18A in May 1957 to test the third stage motor and the Minitrack tracking system. It was a success.

(Below) A Martin technician participates in the checkout of Vanguard TV-2's GE X-405 main engine inside Hangar S at Cape Canaveral in September 1957. The flight with dummy upper stages was successfully completed on October 23, 1957. The small size of the Vanguard vehicle is emphasized in this picture - the first stage was only 45 inches in diameter!

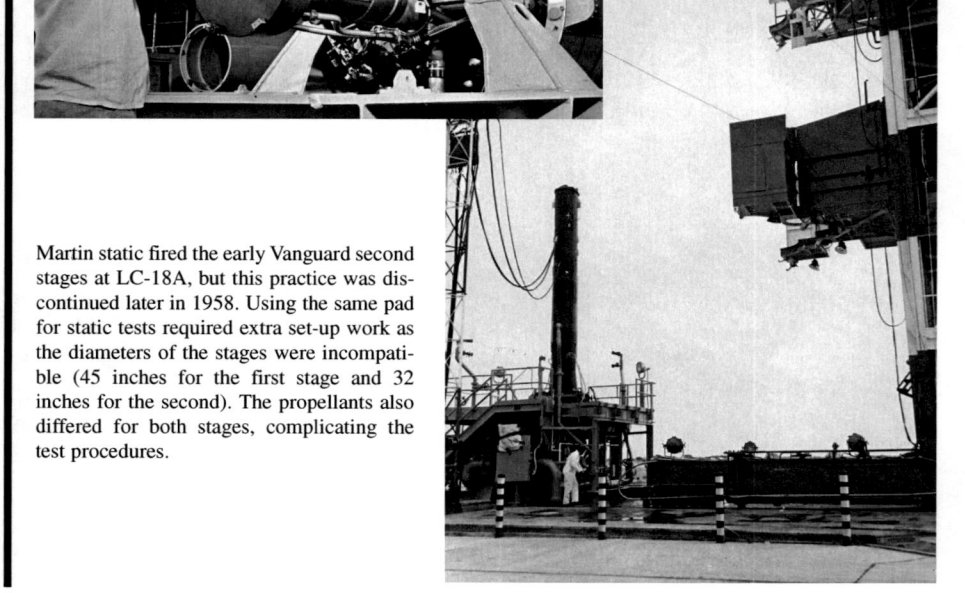

Martin static fired the early Vanguard second stages at LC-18A, but this practice was discontinued later in 1958. Using the same pad for static tests required extra set-up work as the diameters of the stages were incompatible (45 inches for the first stage and 32 inches for the second). The propellants also differed for both stages, complicating the test procedures.

High up in the LC-18A gantry on a cold Florida night two Vanguard 1 personnel check on the satellite aboard TV-4 prior to the successful launch on March 17, 1958.

(Right) Engineers do a fit test for the shroud on an unidentified Vanguard. The shroud was the same size for the early test satellites as it was for the later full sized IGY payloads. Only 3 small test satellites were launched; the remaining eight were the scientific satellites for the 1957-1958 IGY and other experi-ments designed later. The Navaho complex is in the background.

(Left) In early 1958 a loxing Vanguard stands 72 feet above LC-18A against a clear sky during a preflight operations test. The all solid color second stage was a feature of Vanguards TV-3BU (backup) flown February 5, and TV-4 flown on March 17. TV-4 carried Vanguard 1 into orbit, the first of three Vanguard satellites in the eleven vehicle program.

Ground crew conducts a propellant loading exercise for the Vanguard second stage at pad 18. The stage used toxic hypergolic hydrazine and nitric acid to boost the third stage and satellite to orbital injection altitude.

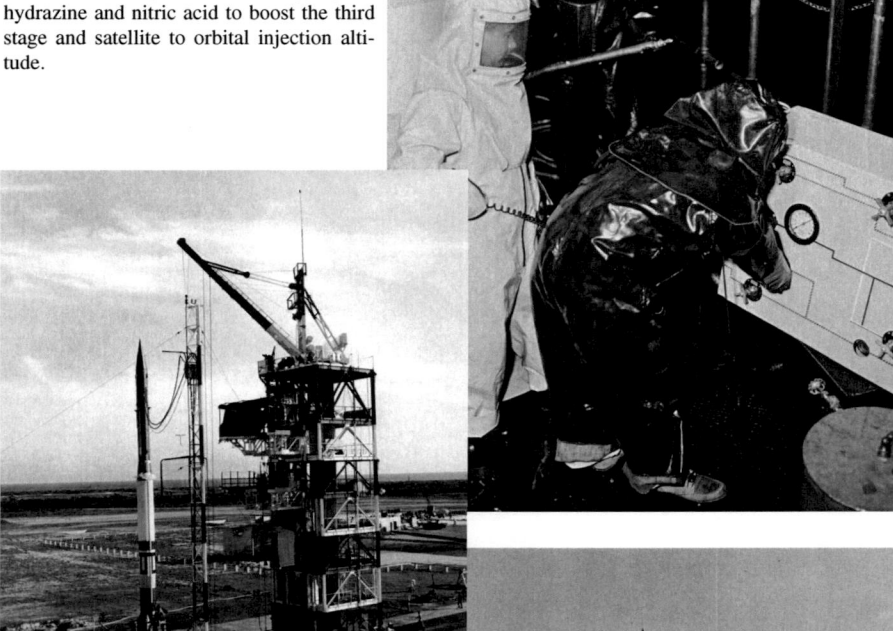

(Above) A high angle view of final preparations to launch the Vanguard 1 satellite on March 17, 1958 at Cape Canaveral. Vanguard vehicle TV-4 is being filled with LOX propellant at pad 18A.

A day to remember: the first successful Vanguard satellite is boosted aloft by TV-4 at 7:15 AM Monday, March 17, 1958 into a sunny, cloud speckled sky. The orbit was more than 400 miles high in perigee confirming the efficiency of the 72 foot tall Vanguard design.

Vanguard SLV-3 would be the last of the NRL flights on September 26, 1958. Other notables for this missile include an engine cutoff 10 days earlier which the vehicle survived intact. The launch was the first to feature the four fall-away arms to allow the engine nozzle more clearance as it rose. A poor performance by the second stage did not provide enough velocity to achieve orbit. Project Vanguard then entered into a stand down until February 1959 after NASA took over on October 1st.

THOR-ABLE
Thor-Able was the first two-stage launch vehicle developed by the U.S. Air Force, combining the Douglas Thor intermediate range missile with the Vanguard second stage rocket. The objective was to test experimental re-entry vehicles at intercontinental distances beginning in early 1958. Thor-Able was quickly adopted by the Air Force to launch the first attempted lunar probes in August, October and November 1958. The 88 foot Thor-Able vehicles used a solid-fuel third stage motor to propel the Pioneer spacecraft to the Moon, but a series of technical mishaps prevented the payloads from reaching their target. The Air Force then utilized Thor-Able for six additional re-entry tests in 1959, at the same time that the new American space agency, NASA, put Thor-Able to work on a series of space launches, including the Explorer 6 Earth satellite, the Pioneer 5 space probe to test long distance interplanetary communications, and the world's first weather satellite, Tiros 1. With a few minor changes the Thor-Able design was transformed by NASA into the Thor-Delta space launch vehicle early 1960.

Never to be forgotten: the first U.S. lunar probe (Pioneer 0) surges aloft from LC-17A at 7:18 am EST on Sunday, August 17, 1958 - just 4 minutes after the launch window opened. A turbopump over-speed caused the vehicle to explode at 77 seconds.

(Right) The second of a three vehicle test series is successfully launched from LC-17 on July 9, 1958 to test an ablative nose cone over intercontinental distances. Efforts to recover the nose cone and its mouse passenger failed. An earlier launch on April 23 also failed.

(Below) A gantry view of the third Thor-Able re-entry test vehicle before its successful flight to the South Atlantic on July 23, 1958. The nose cone survived the intense heat of re-entry but unfortunately was not recovered.

(Right) A new series of six re-entry tests was assigned to Thor-Able for the first half of 1959. A new and advanced RVX-1 nose cone and guidance system were tested. The first flight failed but the remaining five were successful with 2 recoveries being made on the fourth and fifth missions in the program. Thor number 132 boosted this vehicle on March 21, 1959.

Mission number 4 was launched on April 8, 1959 and resulted in the first ever recovery of a nose cone from ICBM range. A prototype Titan guidance system was used to verify the design before Titan was ready for it.

(Below) The Thor-Able for the first U.S. lunar probe is shown venting gaseous oxygen in the predawn darkness of August 17, 1958. The countdown with multiple built-in holds continued as the gantry was pulled back. The new process of an extended countdown allowed for delays without missing the short time frame of a lunar launch window. Later with launch azimuth re-sets and parking orbits the window was extended for Ranger and Lunar Orbiter follow-on missions.

(Left) The second Thor-Able in the first American lunar program was launched on October 11, 1958. The Pioneer 1 space probe reached an altitude of 71,700 miles which was the deepest penetration into space at the time. A minor deviation in pitch prevented the possibility of a lunar flyby. The third and last attempt was launched on November 8, 1958 but a third stage failure ended the USAF/ARPA collaboration without reaching the moon.

(Right) A superb close up of the launch of Explorer 6 with a Thor-Able III rocket on August 6, 1959 from pad 17A (test 1005).

(Below left) As the 6:40 AM launch time approaches for Tiros 1 the gantry is rolled back at pad 17A while additional liquid oxygen replenishment is being done on the first stage. Just after sunrise a perfect launch into a near circular orbit took place. This was the last launch under the Thor-Able banner.

(Below right) As the morning mist clings to the Cape Canaveral horizon, the forerunner of today's world-wide weather forecasting system was launched on April 1, 1960 by the last Thor-Able rocket. Tiros 1 was a spin-stabilized satellite weighing 640 pounds that carried two crude TV cameras. The images were recorded on board Tiros 1 for playback to ground receiving stations.

ATLAS-SCORE

On December 18,1958 the Convair launch crew at Cape Canaveral surprised the whole world by placing an entire Atlas missile into orbit (minus the booster section). The project was kept secret from all but 88 members of the much larger Convair launch team at the Cape, by order of the President. The four ton Atlas 10B vehicle carried an experimental Army Signal Corps space communication package to relay voice and teletype transmissions between several ground stations and the orbiting Atlas. Project SCORE (Signal Communications by Orbiting Relay Equipment) was in effect the world's first communications satellite. The United States achieved an important propaganda victory when the SCORE payload relayed President Eisenhower's Christmas message to the world one day after launch.

(Right) Atlas 10B gleams in the sunlight at Complex 11 on the day that it was launched into orbit on December 18, 1958.

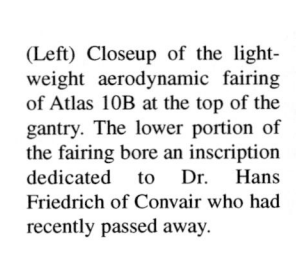

(Left) Closeup of the lightweight aerodynamic fairing of Atlas 10B at the top of the gantry. The lower portion of the fairing bore an inscription dedicated to Dr. Hans Friedrich of Convair who had recently passed away.

ATLAS-ABLE

The least successful of all United States space launch vehicles was the 3 stage Atlas-Able which expended 4 missiles and 3 payloads in a 15 month period (September 1959 to December 1960) which never got anywhere near the moon. The only Atlas C model used in a space role was 9C serving as the booster stage for the first attempt which was destroyed after an abortive static test firing which shut down abruptly after ignition(see Incidents and Accidents section). Failure of the payload shroud during the first flight on Thanksgiving morning in 1959 destroyed the launch vehicle. In 1960 NASA launched two more Atlas-Ables from LC-12 carrying 375 pound Pioneer orbiters, but both failed due to second stage malfunctions.

A promise unfulfilled: the third Atlas-Able points skyward towards its unseen target in December 1960. The goal of lunar orbit would have to wait until 1966 for Lunar Orbiter.

USAF personnel and civilian workers have unloaded the Able stage for the second Atlas-Able mission from a Douglas C-124 Globemaster at the Cape Canaveral Skid Strip on August 22, 1960.

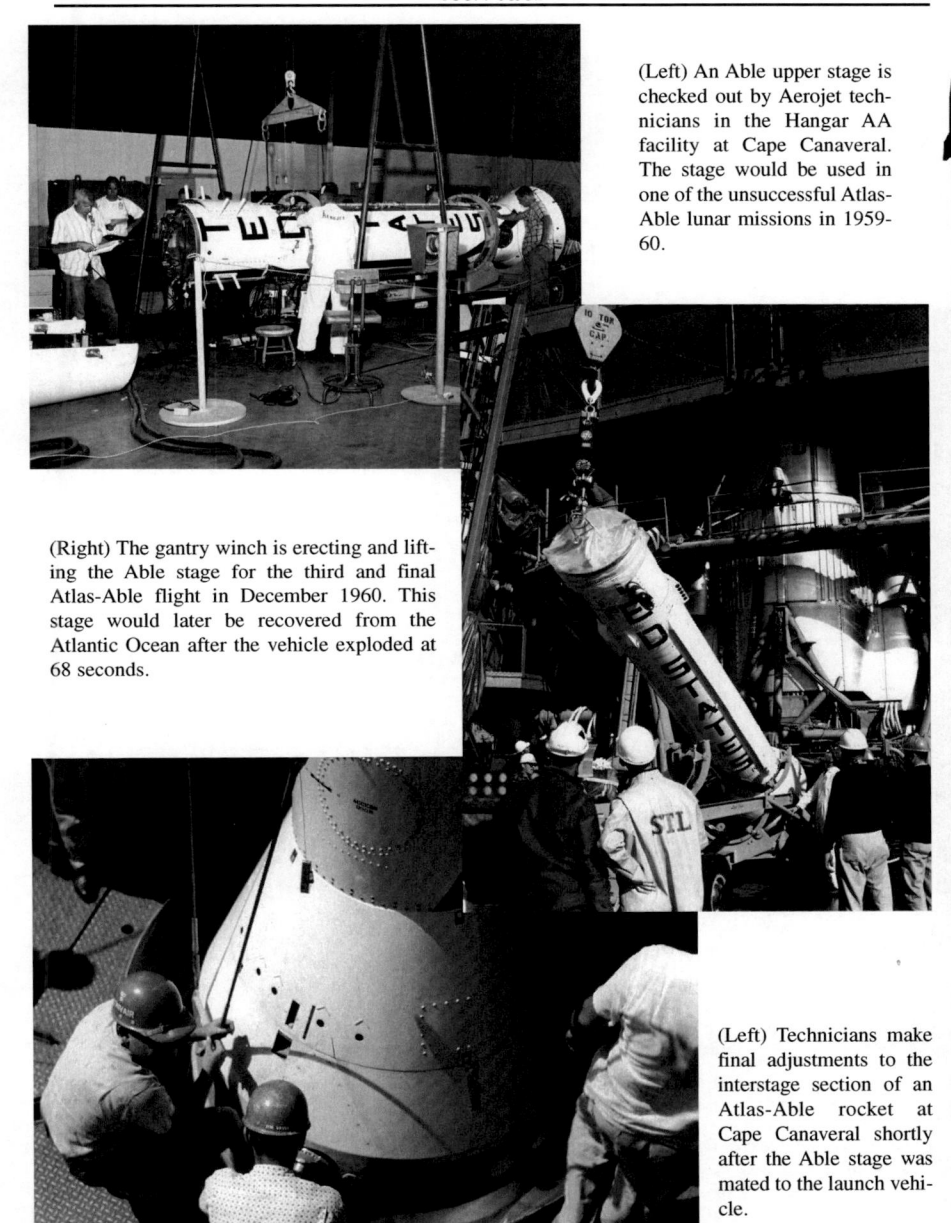

(Left) An Able upper stage is checked out by Aerojet technicians in the Hangar AA facility at Cape Canaveral. The stage would be used in one of the unsuccessful Atlas-Able lunar missions in 1959-60.

(Right) The gantry winch is erecting and lifting the Able stage for the third and final Atlas-Able flight in December 1960. This stage would later be recovered from the Atlantic Ocean after the vehicle exploded at 68 seconds.

(Left) Technicians make final adjustments to the interstage section of an Atlas-Able rocket at Cape Canaveral shortly after the Able stage was mated to the launch vehicle.

(Right) The bulbous payload fairing of an Atlas-Able rocket is lifted to the top of the gantry for mating with the upper stages of the vehicle.

(Below right) The first Atlas-Able poses for the camera as the gantry retracts at LC-12 in September 1959. A static test firing on September 24 resulted in a fire and destruction of the first two stages and damage to the launch pad.

(Above) A ground test of the Atlas-Able shroud explosive separation is conducted with a simulated payload early in the program.

(Left) Poised for flight, the third and last Atlas-Able is photographed from the perimeter road bridge that crossed the exhaust duct water catch basin. Atlas-Agena would replace the Able series in 1961 at the same launch pad equipped with a new umbilical tower.

(Right) Spectacular but ill-fated: the third Atlas-Able vehicle's fiery ascent from pad 14 at Cape Canaveral on December 15,1960. A failure in the Able stage brought the mission to another premature conclusion. It would be four-and-a-half years before the United States would launch a successful flight to the moon.

ATLAS-AGENA

Atlas-Agena first appeared at the Cape under the auspices of the Defense Department's Midas early warning satellite program. Agena was originally developed for the secret Corona satellite reconnaissance program being conducted from the west coast launch site at Vandenberg AFB using Thor rockets. Two flights of the single-burn Lockheed Agena A upper stage were made from Canaveral's Launch Complex 14 in February and May 1960 before the project was moved west to Vandenberg. Next up at the Cape were the NASA Atlas-Agena B and D stages, which had in-orbit restart capability. NASA flew nine Ranger lunar probes on Atlas-Agena in 1961-65, and Atlas-Agena boosted the first planetary exploration missions to Venus and Mars with the early Mariner probes. Agena was NASA's workhorse booster for scientific and space probe missions until Atlas-Centaur became operational in 1966 and assumed the burden of those high priority missions. Atlas-Agena D continued in service at Cape Canaveral (LC-13) until 1978 for various classified military space projects (see "Atlas-Agena at pad 13"). A total of 32 Atlas-Agena launches were conducted for civilian space projects at Cape Canaveral from 1960-1968.

(Right) The first Atlas-Agena D for the Lunar Orbiter program is set up at LC-13 in the summer of 1966. The vehicle number, 17, was painted on the nacelles. Note the liquid nitrogen line installed on the boat-tail just below the vernier engine. Liftoff occurred Successfully on August 10, 1966.

(Below) The first Convair Atlas booster for Atlas-Agena A arrives at LC-14 for the launch of the Midas 1 missile early warning satellite in early 1960. The Atlas phase of the ascent was successful on February 26, 1960 but the Agena stage did not separate and orbit was not achieved. The Agena stage first flew with Thor boosters at Vandenberg AFB in February 1959.

(Right) Gantry rollback during preflight testing at LC-14 on the first Atlas-Agena A. In the early days Atlas umbilical towers and gantries were painted in a colorful red and white candy stripe scheme. The gantry parking area is to the right out of the picture.

(Left) The second Midas launch successful at Cape Canaveral on May 24, 1960. Here the Agena stage is being hoisted past Atlas 45D at LC-14 on its way up to the interstage adapter to be mated with the Atlas. This was the second and last Atlas-Agena A at the Cape.

(Below) The second Ranger deep space orbital attempt is prepared at LC-12 in November 1961. Atlas 117D was used in the unsuccessful launch on November 18,1961 when the Agena B stage failed to fire for the second burn to leave the initial parking orbit. Note the long shroud which follows the shape of the Ranger block I spacecraft. Ranger did not achieve its goal of attitude control trials for its upcoming lunar impact missions in the first two attempts at highly elliptical earth orbits meant to reach out to 675,000 miles.

(Left) The beginnings of the trials and tribulations of the Rangers usually began shortly after the launch for the moon. Here Ranger 4 ignites for its lunar impact mission. The mute spacecraft landed on the far side lunar surface after the master clock failed to open the solar panels and command a sun oriented attitude for the lunar cruise phase. Battery power finally ran down after nine hours of continuous operation.

(Right) With the aftermath of Hurricane Ella showing in the clouds, Ranger 5 is about to discard its environmental cover from the payload fairing as the launcher arms are about to release the Atlas-Agena launch vehicle from LC-12 just after noon on October 18, 1962. A short circuit apparently prevented the solar panels from generating electricity to power the third and last Block II Ranger. The silent lunar probe went into solar orbit after passing the Moon 3 days later.

(Below) A unique sight at the Cape was the removal and re-installation of the sustainer engines of Atlases 121 D (Ranger 3) and 109D (MA-6 flight with John Glenn), seen here at the mouth of the exhaust duct at the pad. This operation was to allow for the removal of the fuel tank insulation by workers entering the open fuel tank and removing the fuel-soaked insulation to prevent it from migrating to the fuel lines under acceleration and entering the engine fuel pumps.

Ready to go: Atlas-Agena and Ranger 7 are poised under the morning sun as they are prepared for the first successful US lunar probe from July 28-31, 1964. Ranger's TV cameras gave scientists the first close views of the lunar surface during the probe's 6,000 MPH descent. This was the 13th United States launch to the moon. Two more Ranger flights ended the program on a high note in early 1965.

Off at 4:21 am on July 22, 1962 from LC-12, the first U.S. Venus probe is boosted aloft atop Atlas 145D and its Agena B upper stage. The Atlas radio guided transponder failed shortly into flight as did the taped backup autopilot program, and when the trajectory deviated from safe limits the range safety officer destroyed the vehicle about 10 seconds before staging at 290 seconds.

(Below) Just days later Atlas 179D is erected at LC-12 as the second vehicle in the two shot series known as Mariner R. On August 27, 1962 Mariner 2 was sent on its way on a historic 109 day voyage to Venus. The flyby December 14, 1962 revealed Venus's high temperatures and crushing air pressure, which were to frustrate early Soviet attempts to parachute to the surface and do scientific experiments.

(Left) Mariner 4's Atlas-Agena undergoes a propellant loading test at LC-12 prior to the final countdown on November 28, 1964. The Mariner 3 spacecraft was launched from LC-13 on November 5, 1964 and was put into a transfer orbit towards Mars. A failure of the special lightweight shroud to jettison trapped the spacecraft inside and eventually the spacecraft failed because it could not deploy the solar panels. A new shroud was designed and manufactured in record time to meet the limited launch window for Mariner 4.

(Right) At 9:22 AM on November 28, 1964 Atlas 288D surges into the cloudy skies above LC-12 to send Mariner 4 on its way towards its famous photographic flyby of Mars on July 14,1965. Mariner 4 recorded 22 images that revealed Mars had impact craters like our Moon.

In 1964 NASA added three earth orbiting scientific satellite programs for Atlas-Agena to launch from LC-12 (although the last flight in March 1968 was transferred to LC-13). Shown is that final launch on March 4, 1968 with Orbiting Geophysical Observatory 5.

(Right) Appearing at first glance to be an Atlas-Centaur, a unique Atlas-Agena undergoes dress rehearsals at pad 12 for the first Orbiting Astronomical Observatory in April 1966. A special interstage adapter and 10 foot diameter shroud were installed on a modified Atlas to carry the payload. A successful launch on April 8 was in vain as the satellite suffered a major power failure after 2 days in orbit and was lost. Future OAO launches were transferred to Atlas-Centaur in 1968.

(Left) Engine start on Atlas SLV-3 #5804 on May 4, 1967 at LC-13 to start the Lunar Orbiter 4 photo recon mission to the moon in support of Apollo. This was the fourth of five successful flights for Lunar Orbiter in its brief history at the Cape from August 1966 to August 1967. The special Kodak film was processed in lunar orbit after exposure and relayed to earth after scanning.

(Above) Last of the planetary flights for Atlas-Agena: Mariner 5 has just lifted from LC-12 on June 14, 1967 for a flyby of Venus using the modified Mariner Mars backup spacecraft, with two less solar panels and new instruments.

(Left) After leaving the Mercury program in 1963, Atlas was employed once again in the manned space flight field to serve as the booster for Gemini target vehicles beginning in 1965. Launch Complex 14 was modified to become the third Atlas-Agena pad at the Cape for the target vehicle launches. Here a NASA Agena stage is about to be mated to the Atlas first stage for a GATV mission.

Under leaden skies one of the six target vehicles awaits its time to fly. No special markings other than a small serial number on the Atlas distinguished these vehicles from each other. The Agena did carry a small logo but the 6 vehicles are not easily identified except by the lighting and weather conditions at launch. Target vehicles flew from October 1965 to November 1966 in support of Gemini missions 6, 8, 9, 10, 11 and 12.

MERCURY-REDSTONE
In preparation for manned orbital flight, NASA modified the U.S. Army's Redstone tactical ballistic missile to carry a Mercury spacecraft with an astronaut aboard on a short 15 minute suborbital trajectory covering about 115 miles in altitude and about 300 miles downrange. The 70 inch Redstone tank diameter was just big enough to attach the Mercury capsule. A total of six Mercury-Redstones went to the launch pad but only 5 took to the skies between December 1960 and July 1961. The first vehicle was replaced after an abortive liftoff. The launch escape tower was triggered after an electrical plug ejected too early, resulting in engine shutdown. The subsequent flight (MR-1 A) was successful but MR-2 in January 1961 subjected chimpanzee Ham to excessive G-forces. A development flight was conducted to validate the fixes in March 1961, ending any chance to beat the Russians into space. . The final two Mercury-Redstones carried astronauts Alan Shepard and Gus Grissom in May and July 1961 on their historic suborbital space flights.

John Glenn and Gus Grissom have a discussion after one of two scrubs for Grissom's MR-4 flight in July 1961.

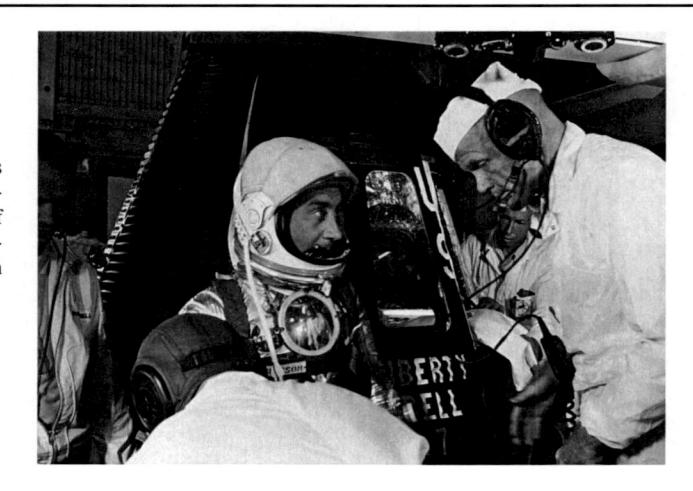

(Left) The first attempt to fly a Mercury capsule aboard a Redstone resulted in an improper umbilical disconnect at liftoff on November 21, 1960. This created an engine shutdown and since the spacecraft had received the engine cutoff signal it promptly initiated a separation and landing sequence. The escape tower is seen just as it left the capsule. To the right the umbilical mast can be seen falling away. The mission of MR-1 was re-scheduled for December as MR-1 A.

(Right) On December 19, 1960 the MR-1A launch took place and was very successful. The spacecraft was recovered 15 minutes later after a 235 mile flight down the Atlantic Missile Raneg with Mercury-Redstone #3 as the booster.

(Left) The next major milestone for Mercury-Redstone was the flight of a 37 pound chimpanzee named Ham to qualify the Mercury spacecraft before an astronaut would climb aboard. Spacecraft Number 5 is lifted to the top of the gantry at pad 5 for mating to the Redstone launch vehicle.

(Right) Liftoff of MR-2 with Mercury-Redstone #2 with Ham aboard. Faulty propellant mixture ratio caused an early engine shut-down which triggered the capsule escape tower to fire and pull the spacecraft towards a higher and longer trajectory. The recovery was made after some delay in reaching the new impact area. Ham was recovered in good condition.

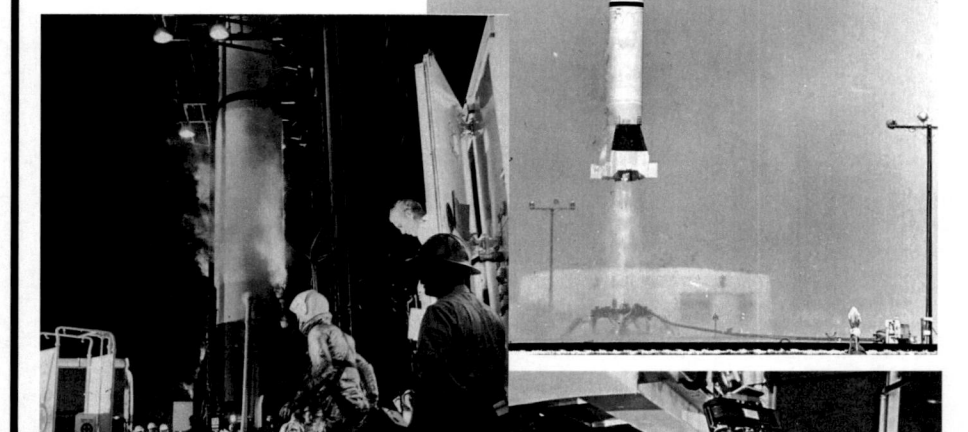

(Above) On May 5, 1961 Alan Shepard steps down from the NASA transport van as he arrives at LC-5 in the pre-dawn darkness for the MR-3 flight. Mercury-Redstone booster #7 is in the background still enclosed by the service tower and fully loaded with liquid oxygen.

(Right) John Glenn, the backup pilot for the MR-3 mission, beckons to Alan Shepard to come over and enter spacecraft Freedom 7. The checkerboard pattern at the top of the Redstone is visible above the floor.

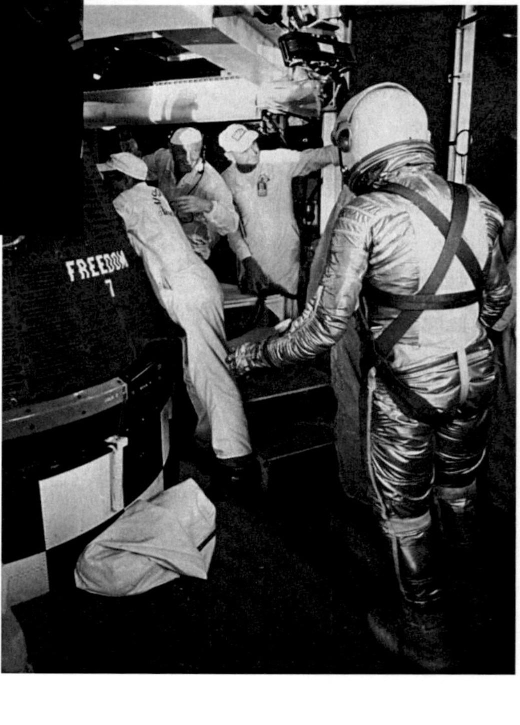

(Left) Well into the countdown MR-3 is being readied for flight shortly after sunrise on May 5, 1961. The former Army gantry has been rolled back to its parked position.

(Below) We are under way. Alan Shepard is propelled into history at 9:34 am EST on May 5, 1961 (test 108) from pad 5 at Cape Canaveral after several delays in the countdown. After 15 minutes and 22 seconds Freedom 7 landed safely and was recovered with astronaut and spacecraft both in good condition. The camera that took this photograph was located at the top of the LC-5 gantry.

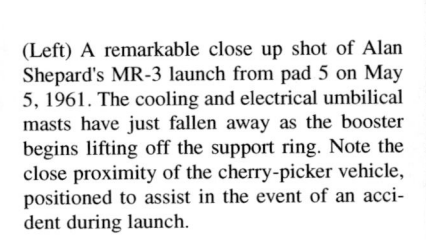

(Left) A remarkable close up shot of Alan Shepard's MR-3 launch from pad 5 on May 5, 1961. The cooling and electrical umbilical masts have just fallen away as the booster begins lifting off the support ring. Note the close proximity of the cherry-picker vehicle, positioned to assist in the event of an accident during launch.

(Right) A Mercury Redstone booster has just arrived at Hangar R for preflight checkouts at the Cape in 1961.

(Below) Mercury Redstone booster number 8 is lifted by the pad 5 gantry crane in preparation for stacking Gus Grissom's MR-4 space vehicle in early July of 1961.

(Right) On the morning of July 19, 1961 Gus Grissom is waiting for launch but the weather caused a second straight postponement after the first attempt on July 18.

On July 21, 1961 MR-4 with Gus Grissom and his Liberty Bell 7 spacecraft is launched on a 118 mile-high trajectory, 303 miles down-range to end the suborbital Mercury-Redstone flights. Grissom nearly drowned when his hatch blew prematurely after splashdown (see Downrange Operations, "The Recovery of Liberty Bell 7.")

ASTRONAUTS AND PRESIDENTS AT THE CAPE

Cape Canaveral became famous all over the world in the late 1950's because of the early Explorer and Pioneer space shots, and for the many missile tests conducted nearly every week at the Florida base. When NASA's Project Mercury arrived on the scene in 1959, the Mercury 7 astronauts became America's newest celebrities, particularly at Cape Canaveral where these men would ride the rockets. Even the chimpanzees that flew the Mercury capsule test flights, Ham and Enos, were briefly captured in the celebrity spotlight at the Cape. President Eisenhower visited the base during his last year in office, and President Kennedy visited Cape Canaveral three times, the final visit just one week before he was struck down by an assassin's bullets in November 1963. The photographs in this section provide just a glimpse of the many astronauts and astro-celebrities who spent some time at Cape Canaveral in the 1960's.

President John F. Kennedy (left) and Vice President Lyndon Johnson (right) confers with NASA astronauts Schirra and Cooper (backs to the camera) inside Hangar S at Cape Canaveral during one of President Kennedy's three visits to the base in 1962-63. NASA Administrator James Webb is visible beside the Vice President.

President Kennedy accompanies Mercury astronaut John Glenn on his welcome home parade on February 23, 1962 at Cocoa Beach, Florida.

(Below) President Eisenhower (center of crowd with hat in hand) visits Launch Complex 14 in January 1960 to see the first Atlas-Agena vehicle (29D). This was Eisenhower's first and only visit to Cape Canaveral.

(Below) Mercury astronauts Deke Slayton (left) and Gordon Cooper (right) inspect pad 13 during one of their early familiarization tours at Cape Canaveral in January 1960.

(Left) Dee O'Hara, the astronaut's nurse at Cape Canaveral, chats with two newly assigned astronaut candidates Elliott See (left) and Ed White in the AeroMed section of Hangar S in October 1962.

(Below) Ham's handlers coax the chimpanzee out of his pressurized flight container during training runs prior to the MR-2 Mercury test flight on January 31, 1961. Ham was successfully recovered despite the fact that his capsule developed a leak at splashdown.

(Below) MA-6 astronaut John Glenn (in pressure suit) is accompanied by Mercury suit technician Joe Schmitt as they leave the crew quarters at Hangar S on the morning of February 20, 1962.

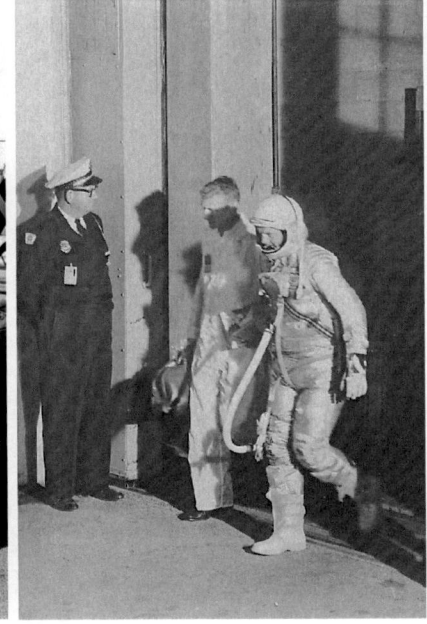

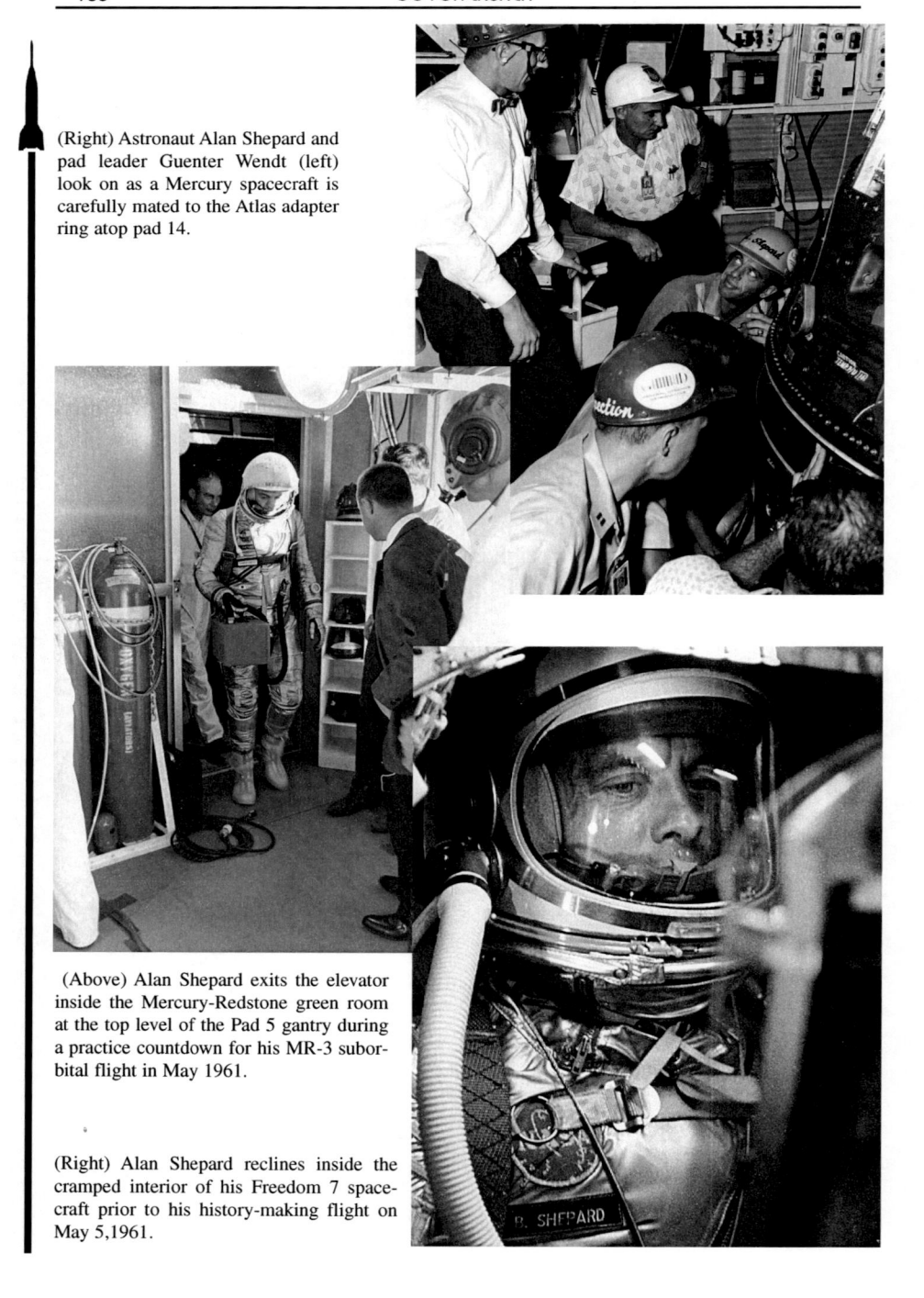

(Right) Astronaut Alan Shepard and pad leader Guenter Wendt (left) look on as a Mercury spacecraft is carefully mated to the Atlas adapter ring atop pad 14.

(Above) Alan Shepard exits the elevator inside the Mercury-Redstone green room at the top level of the Pad 5 gantry during a practice countdown for his MR-3 suborbital flight in May 1961.

(Right) Alan Shepard reclines inside the cramped interior of his Freedom 7 spacecraft prior to his history-making flight on May 5, 1961.

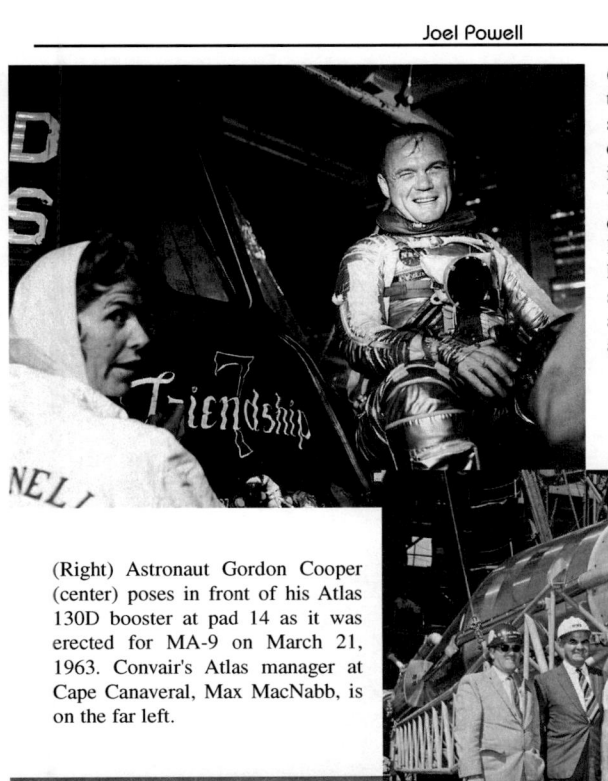

(Left) John Glenn expanded upon the tradition of naming his Mercury spacecraft by designing a more elaborate logo than the block lettering of Freedom 7 and Liberty Bell 7. Glenn asked a graphic artist employed by McDonnell, Cecilia Bibby (seen here touching up her handiwork) to paint the logo on the spacecraft. Bibby also rendered the artwork for Aurora 7 (Carpenter) and Sigma 7 (Schirra).

(Right) Astronaut Gordon Cooper (center) poses in front of his Atlas 130D booster at pad 14 as it was erected for MA-9 on March 21, 1963. Convair's Atlas manager at Cape Canaveral, Max MacNabb, is on the far left.

(Left) Cooper exits the transfer van at pad 14 on launch morning (May 15, 1963) as a crowd of space workers gather to salute the final flight of Project Mercury.

President Kennedy (center) attends a Saturn rocket briefing at the LC-34 blockhouse in September 11, 1962 with NASA Administrator Webb (far left), Vice-President Johnson, NASA Center Director Kurt Debus and Defense Secretary McNamara (far right).

An ebullient Wernher von Braun in his natural element- in this case the LC-37 blockhouse during the countdown of Saturn I SA-5 in January 1964.

MERCURY-ATLAS
The Atlas ICBM became an integral part of the U.S. manned space program when it was selected for Project Mercury in 1959. A major modification was to install an abort sensing system (ASIS), plus an adapter to mount the 3,000 pound Mercury spacecraft. Ten Atlas D models were flown in the program from September 9, 1959 to May 15, 1963 at Launch Complex 14 (where NASA added an emergency astronaut escape ramp). Four of these Atlas boosters carried American astronauts into Earth orbit. Three of the first four unmanned test flights malfunctioned, but only one design change was instituted: an increase in the skin thickness of the Atlas structure directly beneath the Mercury spacecraft adapter. The third failure validated the spacecraft escape system when Range Safety triggered the engine cutoff command to initiate an abort. The escape rocket pulled the capsule away from the malfunctioning Atlas, which was recovered without damage and was refurbished to fly the next test flight. Mercury-Atlas was the only expendable American launch vehicle capable of placing a spacecraft into low-Earth orbit without an upper stage.

(Left) Mobile spotlights illuminate Launch Complex 14 prior to the launch of MA-9 on May 15, 1963. Gordon Cooper flew a 22 orbit mission which completed the Mercury program. All four Mercury-Atlas manned flights were successful in a period of just 15 months.

(Right) The first boilerplate Mercury spacecraft is being readied for hoisting aboard Atlas 10D at LC-14. The number 628 is the last 3 digits of the USAF serial number which has replaced the Convair model number, usually stenciled on the tank instead.

(Left) A view of LC-14 with service tower retracted to park position and umbilicals connected to Atlas 10D. The first Mercury test flight on September 9, 1959 was highly successful and the spacecraft recovered despite its failure to cleanly separate after sustainer engine cutoff. It re-entered the atmosphere with the heat shield initially pointing away from the re-entry path. The off axis center of gravity oriented the heat shield after the depleted attitude control jets had stopped thrusting.

(Right) A static test firing is conducted on Atlas 50D in the summer of 1960 at LC-14, prior to its failed flight on July 29, 1960. The vehicle broke up at 38 seconds up due to severe stresses at the spacecraft adapter and lox tank interface. This was the last spacecraft to fly without an escape tower and abort system.

(Above) Atlas 67D shows its adapter band to strengthen its thin skinned lox tank area near the capsule adapter. Later Atlases would have a thicker skin so the band was used only once for the MA-2 light on February 21, 1961. The spacecraft was recovered after its ballistic flight 1,244 nautical miles downrange from Cape Canaveral.

(Right) A Mercury-Atlas launch vehicle arrives at the Cape Canaveral Skid Strip and has just been unloaded from the Douglas C-133 Cargomaster aircraft.

(Left) Atlas 100D is backed up a ramp leading to the pad at LC-14 in March 1961 in preparation for the first orbital launch for Mercury on the MA-3 mission. The launch on April 25 was in immediate trouble as 100D climbed up past the umbilical tower at the two second mark and the roll and pitch program did not start. At 42 seconds the RSO sent the engine cutoff command to the vertically ascending Atlas, which triggered the abort system to fire the escape tower. Atlas 100D was destroyed as spacecraft number 8 was carried to 24,000 feet altitude by the escape tower before parachuting gently onto the nearby beach.

(Above) Atlas 88D arrives of LC-14 for the MA-4 flight which flew the first successful orbital mission in Project Mercury on September 13, 1961. Spacecraft # 8A (which had originally flown on the failed MA-3 shot) was recovered after a single earth orbit.

(Left) A Mercury-Atlas rocket is enclosed within the work platforms of the LC-14 gantry (seen from the uppermost level) shortly after being installed in the launcher mechanism.

A fine portrait of the Mercury-Atlas launch facilities at Complex 14 taken from the top of the service tower. The blockhouse with its four periscopes for viewing the launch is visible in the background. Atlas 100D was aborted shortly after liftoff on April 25, 1961 during the first attempt to place an unmanned Mercury spacecraft into orbit.

(Left) Mercury spacecraft reaches the upper level of the pad 14 gantry prior to being mated to the waiting Atlas booster (foreground).

(Below) The Mercury-Atlas 5 spacecraft, which made a shortened two orbit flight with the chimpanzee Enos is about to be mated to Atlas 93D in the pad 14 white room. The flight on November 29, 1961 paved the way for the first manned orbital space shot with John Glenn on February 20, 1962.

(Left) The last step in the assembly of a Mercury-Atlas vehicle at Complex 14 was to place the escape tower on top of the Mercury spacecraft. The solid rocket motor was designed to pull the Mercury capsule and astronaut to safety in the event of a malfunction in the Atlas booster.

(Right) Atlas 100D is shown during gantry rollback at LC-14 prior to the ill-fated Mercury-Atlas 3 flight. This was the first of the thicker skinned Atlases that could withstand the stresses of carrying the heavy Mercury spacecraft. See "Incidents and Accidents" section for the outcome of this flight.

(Below) Thundering into a cloudy morning sky, Atlas 88D/ MA-4 takes the first orbital Mercury spacecraft (#8A) towards a 28.5 degree orbit around the earth on September 13, 1961. This mission opened the door to fly Enos into orbit in November.

(Right) During preparations for Glenn's flight it was discovered that the fuel tank insulation membrane had leaked, allowing the insulation to become soaked with the RP-1 kerosene derivative fuel. To remove the insulation, engineers dismantled Atlas 109D's sustainer engine and the fuel tank apex cones which allowed technicians to climb into the tank with make-shift scaffolding and remove the insulation.

(Right) An intimate fish-eye portrait of John Glenn entering his Friendship 7 Mercury capsule in the white room of pad 14 on the early morning of February 20, 1962.

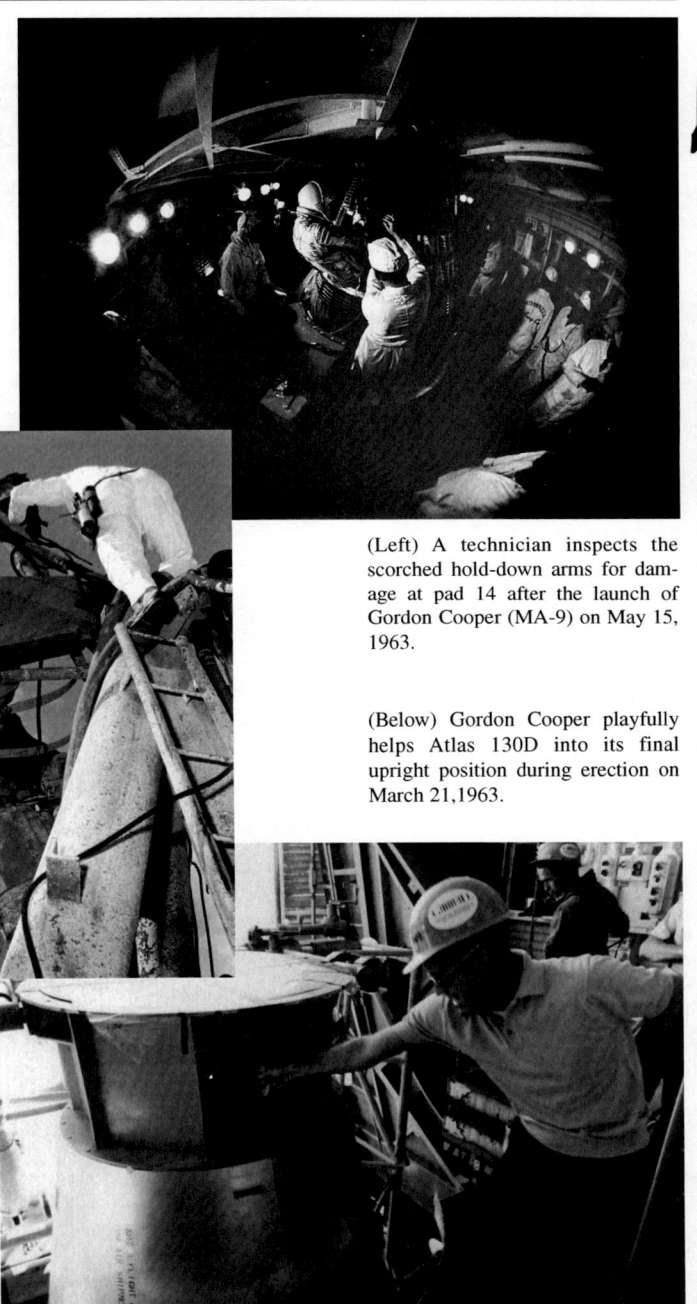

(Left) A technician inspects the scorched hold-down arms for damage at pad 14 after the launch of Gordon Cooper (MA-9) on May 15, 1963.

(Below) Gordon Cooper playfully helps Atlas 130D into its final upright position during erection on March 21, 1963.

Astronaut Cooper is seen at the spacecraft level in the LC-14 gantry during practice runs for the upcoming MA-9 mission, which would be the last Mercury flight.

Atlas 130D roars aloft from pad 14 on May 15, 1963 carrying astronaut Gordon Cooper on the final flight of Project Mercury.

THOR-STAR-ABLE
A new concept in satellite launching was opened by the ARPA/USAF when the new Aerojet-General Able-Star stage made its first flight as a Thor-Able-Star two stage missile in April 1960. The restartable engine and the larger diameter second stage allowed for near circular orbits using only a 2 stage configuration and was first used for Transit 1B (the second in the series of a new navigation program from space). The triangular first stage fins were flown only on the first launch. Eleven Thor-Able-Star vehicles were flown from Cape Canaveral before the last eight were transferred to Vandenberg AFB for polar orbit missions. This was the fourth major use of the Thor IRBM as a first stage in the evolving US missile and space programs. Major satellite achievements included Transit, Courier and ANNA, and the novel orbiting of 2- and 3- satellite clusters. Thor-Able-Star flights were conducted from LC-17A and B from 1960 through 1962.

Predawn scene at LC-17A for first Thor-Able-Star launch. Transit 1B was the payload for the first ever second stage engine re-ignition in space.

(Right) Using the service tower as a lifting device for the Able-Star stage for Courier 1A the ground crew prepares to raise the rocket high enough to mate with the Douglas first stage Thor at LC-17B.

(Left) The ground crew prepares to hoist the Able-Star stage for the Courier 1A US Army communication satellite launch in August 1960. This first attempt failed in flight.

Transit 3A is unsuccessfully launched on November 30, 1960 from LC-17B along with 3 smaller satellites by Thor-Able-Star. A first stage failure ended the flight.

A view of the first Thor-Able-Star from the partially retracted tower at LC-17B. Thor booster 257 was used for the successful flight on April 13, 1960.

The Thor-Able-Star stage is carefully moved into position for mechanical and electrical mating with the Thor booster prior to flight on August 20, 1960. A hydraulics failure in Thor # 262 caused the vehicle to break up just before the end of the Thor's powered flight.

ANNA 1 B sits enshrouded atop the eleventh and final Thor-Able-Star to fly from Cape Canaveral on October 31, 1962. The success/failure record was 7 and 4 at the Cape while the last flights at Vandenberg AFB with polar orbiting Transit satellites yielded a 7 -1 record.

THOR-DELTA and DELTA

In 1960 an uprated flight controller in the second stage produced Thor-Delta from the proven Thor-Able launcher for NASA. The first mission on May 13, 1960 was a failure, but the second flight was successful with the famous Echo balloon satellite, starting a string of 22 consecutive successes before vehicle 24 failed in August 1964. Improvements such as an increase in the thrust of the first stage from 150,000 to 110,000 pounds and a longer second stage were introduced to gradually increase payload capacity to low earth orbit from 286 pounds in 1960 to over 1,400 pounds by 1971. The next major change was the addition of three solid rockets to boost heavier payloads to geosynchronous orbit, such as Syncom and Early Bird. The next upgrade was to increase the diameter of the second stage (using the 65 inch Able-Star tankage) to produce the Delta E in 1965. Eventually the upper tank section of the first stage was increased to a constant 8 feet diameter, and additional solid motors (up to a total of nine) were added. When an interstage structure was introduced to bring the diameter of the upper stages to 8 feet- the smaller diameter second stage was nestled down inside the interstage cylinder - this became known as the "Straight 8" configuration (2000 series). The Straight 8 became the standard external configuration of Delta in 1974 and continues as such to the present day. After 28 years of exemplary service, the Delta was replaced by the current Delta II vehicle in 1989. This is quite an accomplishment for a launch vehicle that was intended only as a temporary stop-gap booster back in 1960!

Delta 25 with solid rocket motors is poised for launch at pad 17A in August 1964. Note the newly re-built vehicle mounting structure on the pad surface that replaced the familiar launch ring used since the first Thor launches in 1957.

(Right) Thor booster 295 is being trucked to LC-17 for the launch of Explorer 10 by Thor-Delta 4. Hangar M was used by Douglas Aircraft Company pre-flight checkout of its missiles after air transport from Santa Monica, California.

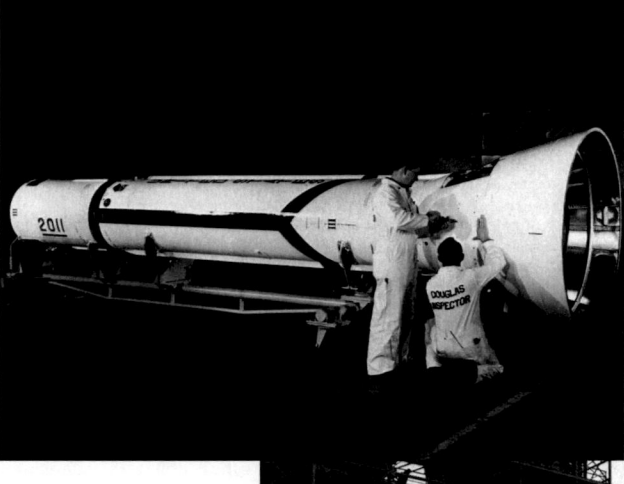

(Left) Douglas inspectors check out the interstage unit of an early Thor-Delta second stage at the Douglas hangar at Cape Canaveral.

(Right) On March 3, 1961 the Thor first stage for Thor-Delta 4 arrives at the launch site. The Thor was hoisted off the truck for installation in the gantry, which was used to move the rocket to the launch pad for mounting on the circular firing table at LC-17B. Delta 4 was launched successfully on March 25, 1961.

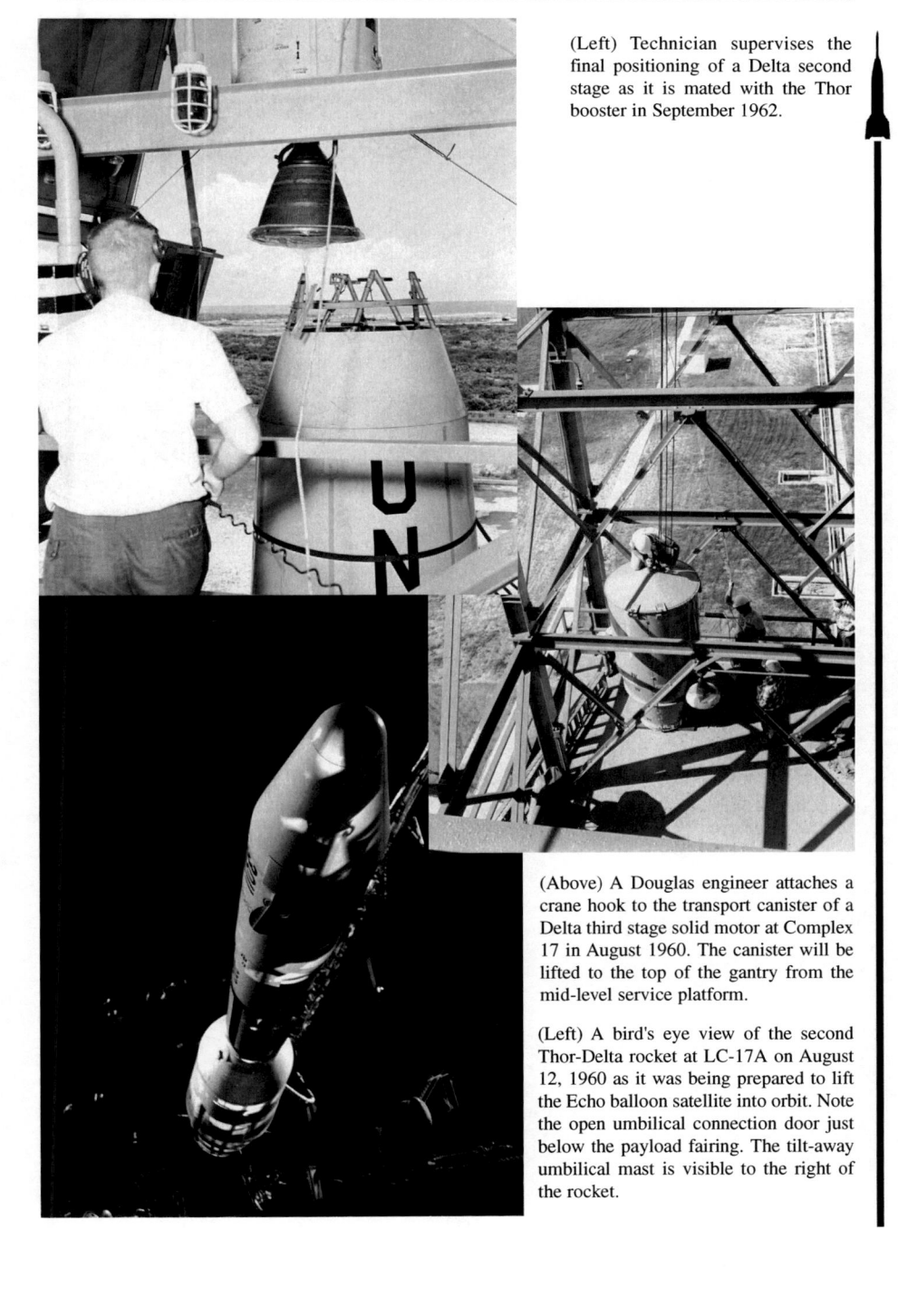

(Left) Technician supervises the final positioning of a Delta second stage as it is mated with the Thor booster in September 1962.

(Above) A Douglas engineer attaches a crane hook to the transport canister of a Delta third stage solid motor at Complex 17 in August 1960. The canister will be lifted to the top of the gantry from the mid-level service platform.

(Left) A bird's eye view of the second Thor-Delta rocket at LC-17A on August 12, 1960 as it was being prepared to lift the Echo balloon satellite into orbit. Note the open umbilical connection door just below the payload fairing. The tilt-away umbilical mast is visible to the right of the rocket.

(Right) Delta 18 was an early version of Delta that lofted the experimental Telstar 2 communication satellite from pad 17A in 1963.

(Below) The very first Intelsat communications satellite was known as Early Bird and was launched into orbit aboard Thor- Delta 30 (the second and last D model). The use of the 3 strap on solid motors allowed this innovative satellite to be placed into a geosynchronous orbit where it appeared stationary relative to the ground, from which it could receive and transmit communications traffic. This historic launch took place on April 6, 1965 from LC-17A.

(Right) Delta E number 67 is shown on LC-17B from beneath the retracted gantry. This uprated version would carry the ESSA 9 weather satellite into orbit on February 26,1969. Note the newer and larger service tower compared to the scene 3-1/2 years earlier with Delta 30 (above).

(Left) A beautiful scene of Delta 67 illuminated by searchlights in February 1969. The triangular Delta logo with the number 67 is quite striking. The three Thiokol Castor solids gave a 38 second push aloft helping overcome the inertia of the heavier E vehicles. All 23 Delta E's were successful in orbiting their payloads.

(Below) Closeup of the Delta 83 vehicle standing on pad 17A during a radio frequency interference test in March 1971. The payload was the IMP-9 science satellite. This is an Delta N variant.

(Left) An advanced, all white Delta 1914 is shown on LC-17B with its enlarged second stage and shroud and nine Castor 2 strap on solids prior to launching the ANIK communications satellite on April 20, 1973. Delta 94 was a "Straight 8" configuration and bore little resemblance to the early Deltas. Note the umbilical tower has been enclosed against the elements.

Technicians get ready to mount a Castor solid rocket motor in the gantry at pad 17A prior to moving the motor to the rocket for mating with the Delta first stage.

Delta 100 is ready for launch with the British Skynet IIA satcom on January 18, 1974 at Pad 17B. The payload failed to achieve geosynchronous orbit. Note the dark color (green) of the 2000 series vehicles, which is standard to this day.

High angle view of two Delta solid rocket motors inside the gantry (left) before being mated to the Delta first stage. The date was September 1986.

(Right) The Aerojet second stage for Delta 179 is hoisted by crane to be mated with the vehicle (lowered down into the interstage structure sitting atop the Delta first stage).

A fine study of the Delta 121 vehicle at pad 17A in March 1976 with the SATCOM-2 payload. This is one of the final Delta versions (3914) with the larger Castor 4 solid motors.

Delta 179's eight-foot-diameter interstage section is lifted from the delivery truck of pad 17A to be mated with the Delta first stage. The second stage of the rocket is positioned inside the interstage ring for launch.

ATLAS-CENTAUR

NASA's Centaur was the first hydrogen-powered upper stage in the world. Combined with the Atlas missile as a first stage, Atlas-Centaur was the most efficient launch vehicle ever flown thanks to the hydrogen and oxygen propellants, which is one of the most energetic chemical rocket fuel combinations available. Atlas-Centaur was NASA's second attempt to develop a space launch vehicle (the first was the ill-fated Atlas Vega program that was cancelled in December 1959). The first Atlas-Centaur launch attempt ended in disaster on May 8, 1962 (see Incidents and Accidents section) and NASA spent the next four years re-engineering and testing the rocket before it was certified to carry operational payloads in May 1966. Atlas-Centaur soon became one of the workhorses of the civilian space program, launching NASA's largest scientific payloads from Cape Canaveral, and propelling most of NASA's space probes to Mars, Venus, Mercury and the outer planets from 1969 to 1978. From 1978 to its retirement in February 2005, Atlas-Centaur was employed in the commercial launch business to orbit communication satellites for clients all over the world.

(Left) A high-level view of Atlas-Centaur 44 at pad 36A as the service structure is retracted prior to launch on February 9, 1978. Atlas-Centaur was the mainstay of NASA's launch vehicle fleet for middle-weight missions from 1966 to 1989, when the manufacturer assumed control of the vehicle for commercial space launches until the Atlas III model was retired in February 2005.

(Below Left) The Centaur stage has been hoisted inside the service structure at LC-36 and is painstakingly lowered onto the interstage adapter (right) for mating to the Atlas.

(Left) The Centaur stage is lowered to the interstage adapter for mating with the Atlas. The fiberglass insulation panels were attached to the sides of the Centaur after mating was completed.

(Below) For the early Centaur test flights, the two halves of the aerodynamic fairing were installed from inside the service structure.

(Left) The completed fairing for the first Centaur flight, with its pointed aerospike fixture on top. Only the first three Centaur vehicles featured an aerospike on the top of the fairing with an angle-of-attack indicator.

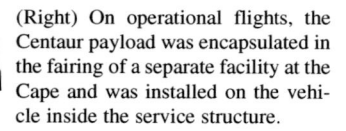

(Right) On operational flights, the Centaur payload was encapsulated in the fairing of a separate facility at the Cape and was installed on the vehicle inside the service structure.

(Above) A plumber's nightmare of umbilical cables and propellant fill-and-drain flex hoses were attached to the vehicle interstage area. All the umbilical connections are visible (right) at pad 36A with the AC-7 vehicle (Surveyor 2, September 1966).

Two different views of the first Atlas-Centaur vehicle (F-1) at launch Complex 36A in May 1962 and one of AC-3 in June 1964. Note the diamond-shaped, angled flame duct. The blockhouse looms in the distance at right in the third picture below.

A marvelous portrait of the Atlas-Centaur launch controllers inside the LC-36 blockhouse on the afternoon of May 8, 1962 on the occasion of the first Centaur launch (F-1).

(Below) The service structure has been pulled back and chilldown of the Atlas propellant tanks has just begun at pad 36B in June 1972. Note the hip young engineer with the Dilbert tie strolling nonchalantly down the ramp. The payload on that day was an Intelsat IV satcom.

(Below) AC-47 has just lifted off from pad 36A on May 4, 1979 in this view looking up the service ramp. Notice the stand of trees behind the gantry serving as a wind break from the direction of the ocean.

TITAN IIIA

The Air Force's Standard Launch Vehicle - 5 (SLV-5) in the 1960's consisted of the Titan IIIC heavy lift launcher (SLV-5C) and the Titan IIIA (SLV-5A), which was the prototype of the Titan IIIC core stage minus the two large solid rocket boosters. Titan IIIA was flown four times from Complex 20 to test the new upper stage for Titan IIIC, the bi-propellant Transtage. The maiden launch of Titan IIIA was on September 1, 1964 (a failure of the Transtage) and the final flight was on May 6, 1965, just over one month prior to the first launch of the massive Titan IIIC vehicle from Complex 40 (see ITL section). At 126 feet in length, Titan IIIA was nearly identical in configuration to the Titan IIIC core stage, both of which were directly based upon the Titan II ICBM.

Liftoff of a Titan IIIA launch vehicle on an orbital test flight from Complex 20 in 1965.

TITAN IIIC

Titan IIIC was a heavy-lift booster developed by the U.S. Air Force in the 1960's to launch large military payloads such as communications satellites. A later version of Titan III was intended to launch the Manned Orbiting Laboratory (MOL) from Vandenberg AFB. The Titan IIIC was one of the most unusual launch vehicles to be developed for the American space program. The "core" stages were conventional enough, derived directly from the Titan II ICBM, but the primary motive power for Titan IIIC was two huge solid rocket motors 112 feet long and ten feet in diameter. Designated Stage Zero, the solid motors burned exclusively for the first minute and 51 seconds until burnout, at which point the first Titan core stage was ignited, and the three Titan stages completed the ascent to orbit. The thirty-eighth and final Titan IIIC was launched from pad 40 of the Integrate, Transfer and Launch facility on March 6, 1982. The replacement for Titan IIIC was called Titan 34D, which featured larger solid rocket motors and an elongated core stage (see Titan 34D chapter at the end of Part 2). NASA's Titan IIIE Centaur version was a basic Titan IIID vehicle (no Transtage) used at Vandenberg AFB, adapted to use the Centaur stage to boost large payloads to the planets.

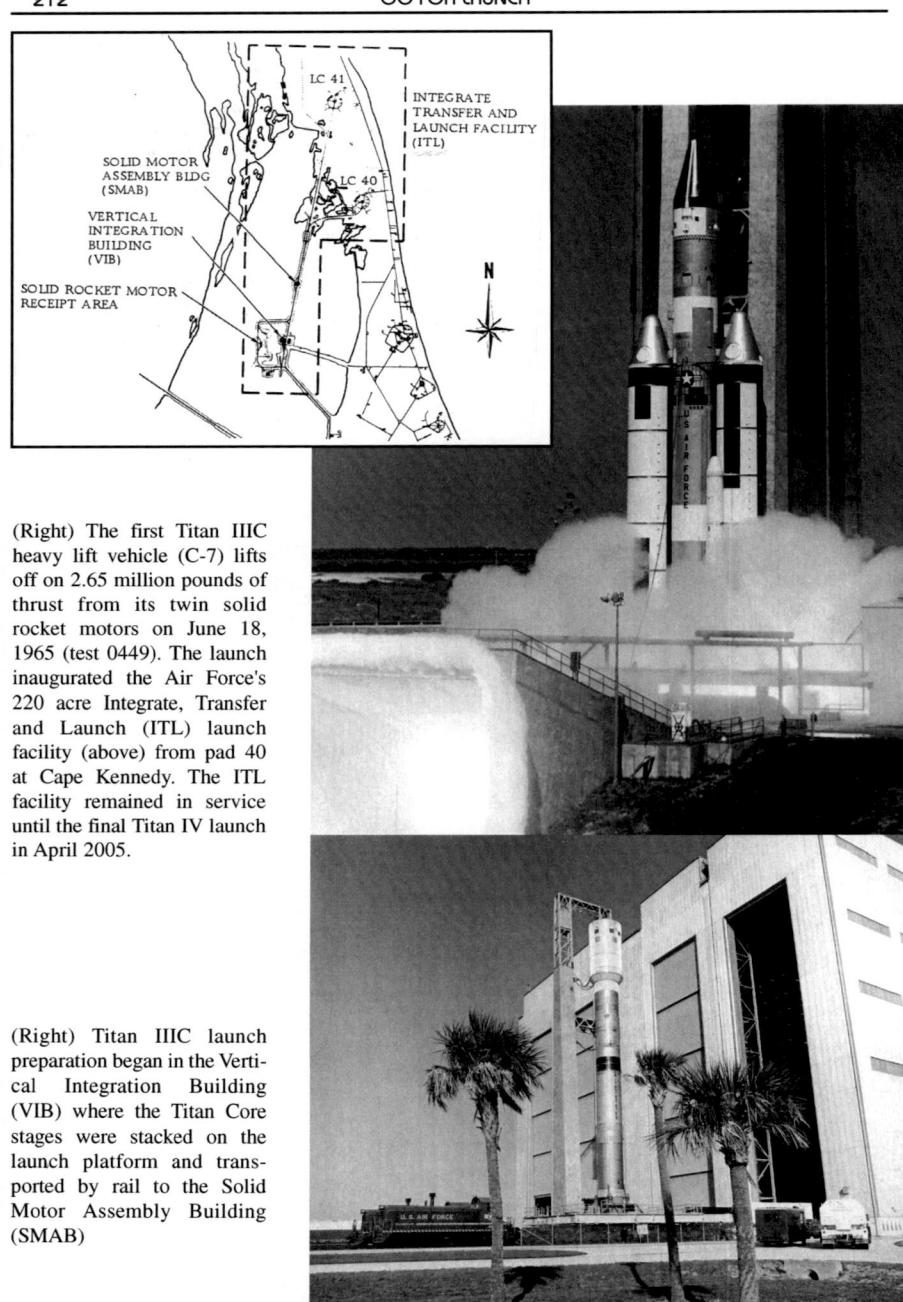

(Right) The first Titan IIIC heavy lift vehicle (C-7) lifts off on 2.65 million pounds of thrust from its twin solid rocket motors on June 18, 1965 (test 0449). The launch inaugurated the Air Force's 220 acre Integrate, Transfer and Launch (ITL) launch facility (above) from pad 40 at Cape Kennedy. The ITL facility remained in service until the final Titan IV launch in April 2005.

(Right) Titan IIIC launch preparation began in the Vertical Integration Building (VIB) where the Titan Core stages were stacked on the launch platform and transported by rail to the Solid Motor Assembly Building (SMAB)

For mating with the solid rocket boosters and the Transtage upper stage and payload. The completed Titan stack was then transported to pad 40 or 41 (the Titan IIIE vehicle is depicted in the photos).

The burned out booster of the first Titan IIIC vehicle begins to tumble as it drops back to the Atlantic Ocean in this frame from an on-board motion picture camera. Cape Kennedy (Canaveral) can be glimpsed to the right of the expended booster.

(Left) The right hand solid rocket motor is attached to the Titan III core stage inside the SMAB facility.

(Right) The 135 foot Titan IIIC rocket towers over the photographer at the base of the launch pad. SLV-5 is the Air Force designation for the Titan IIIC launch vehicle. The second cylinder on the right is the Thrust Vector Control (TVC) fluid reservoir that is used to deflect the exhaust plume to steer the rocket in flight without having to move the nozzles.

A different point of view showing the first Titan IIIC vehicle at pad 40 in June 1965. The 170 foot umbilical tower is at left and the Mobile Service Structure is in the background.

(Right) Titan IIIC facilities include the Umbilical Tower (left) and the massive Mobile Service Structure on the right, present at both ITL pads.

(Left) The first Titan IIIC vehicle accelerates towards orbit as seen from the Titan II facility at Complex 15 south of the ITL site.

A generation removed: From nearly identical vantage points at neighboring Kennedy Space Center, these Titan IIIC launches are separated by nearly 16 years on the calendar and a whole generation at the launch pad. Taken on June 16, 1966, the photo below shows a Titan launch from pad 41 with the Saturn V facilities vehicle at pad 39A, while the photo above shows Titan IIIC-20 lifting off from nearby pad 40 on March 6, 1982 with the shuttle Columbia (STS-3) on the same pad 39A (modified for the Space Shuttle).

(Right) An unusual side view of a Titan IIIC Launch (C-32) on May 12, 1977 from pad 40. The ITL processing facilities are visible across the Banana River.

NASA's mighty Titan IIIE Centaur launch vehicle at Cape Canaveral's pad 41 in the late 1970's.

(Right) The many umbilical cables and propellant replenishment hoses that connected the cryogenic Centaur stage with the Titan IIIE launch vehicle at pad 41.

Joel Powell 217

(Left) With the Centaur stage on top, the Titan IIIE was a much larger rocket than the Titan IIIC: 158 feet vs. 145 feet long. This particular vehicle carried the Viking 1 space probe to Mars on August 20, 1975 from LC-41.

(Below) When the Centaur stage was utilized for NASA interplanetary missions in the 1970's, the encapsulated payload was added to the rocket at the launch pad.

(Below) Another Titan Centaur launch seen from across the Banana River emphasizes the wetlands character of the northern end of Cape Canaveral.

GEMINI-TITAN AT COMPLEX 19

Launch Complex 19 was the site of ten manned Gemini launches from March 1965 to November 1966. American astronauts learned to walk in space, rendezvous and dock with orbiting targets, and to survive up to two weeks in space after launching from pad 19 on modified Titan II missiles. The astronauts were recovered in the Atlantic Ocean downrange from the Cape (except for the emergency landing of Gemini 8 in the Pacific). LC-19 was a former Titan I facility (last use was on January 29, 1962) that was rebuilt for Gemini by NASA from September 1962 to August 1963. The facility retained the Titan erector-gantry system, with the addition of a 50 foot "white room" enclosure at the top of the erector-gantry to shelter the Gemini spacecraft during processing activities. The complex was activated by NASA on October 17, 1963. A short static test tower was also built for the Titan's second stage, and a dual static firing of both stages was successfully conducted on January 16, 1964 with the Gemini-Titan 1 vehicle.

The 138 foot long erector gantry at pad 19 is slowly being lowered to the ground (parked position) during systems checks prior to the first Gemini flight (GT-1) in January 1964.

(Right) The first stage of a Gemini-Titan II launch vehicle is backed into position inside the erector gantry in preparation to raise the rocket onto the launch stand at pad 19.

(Left) The Gemini-Titan second stage, inside the erector gantry, is slowly moved into position for mating with the first stage at pad 19 (visible in the background).

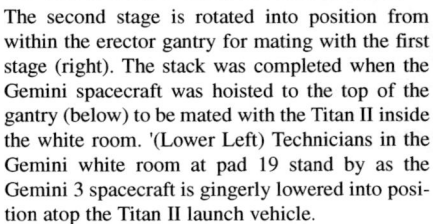

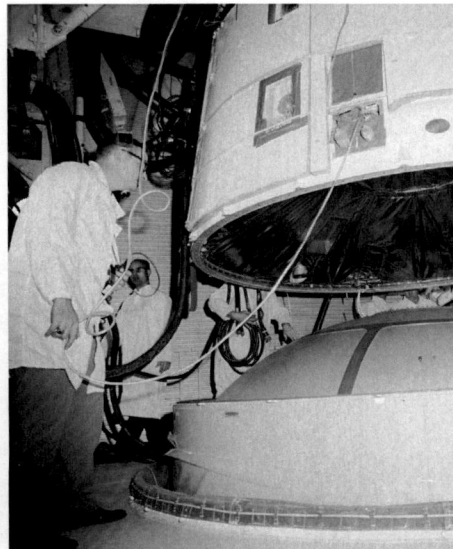

The second stage is rotated into position from within the erector gantry for mating with the first stage (right). The stack was completed when the Gemini spacecraft was hoisted to the top of the gantry (below) to be mated with the Titan II inside the white room. '(Lower Left) Technicians in the Gemini white room at pad 19 stand by as the Gemini 3 spacecraft is gingerly lowered into position atop the Titan II launch vehicle.

The Gemini 6A astronauts stride up the ramp to the elevator at pad 19 to board their spacecraft on December 15, 1965.

(Left) Gemini 6 astronauts Tom Stafford (left) and Wally Schirra (right) are pictured inside the suit-up trailer before launch. The trailer was located at pad 16 (adjacent to Gemini's pad 19) for the convenience of the flight crews.

(Right) With the erector gantry retracted to the ground park position, the Gemini-Titan II launch vehicle is ready for flight after final checkout and propellant loading.

(Left) With the closeout crew about to evacuate the pad 19 white room prior to the terminal countdown for Gemini 10 on July 18, 1966, a camera with a fish-eye lens records the scene in the open-air facility.

(Right) Gemini 8 astronauts Dave Scott (left) and Neil Armstrong are secured in their seats prior to hatch closure in the pad 19 white room on March 16, 1966.

(Left) Launch controllers sit at their consoles inside the crowded pad 19 blockhouse during a Gemini countdown.

(Right) Submarine-style periscopes provide controllers with a view of the launch from inside the heavily fortified concrete blockhouse at Complex 19.

(Left) Liftoff! Another countdown is successfully concluded as the Gemini 11 launch is captured by a remote camera inside the erector gantry.

(Right) The liftoff of Gemini 9A on June 3, 1966 from pad 19, one of the most unusual and visually exciting launch photos ever captured on film at Cape Canaveral.

SATURN I and IB

The first U.S. heavy booster was Wernher von Braun's Saturn I rocket, developed by the Army Ballistic Missile Agency (Huntsville, Alabama) beginning in 1958. The Saturn concept involved clustering existing Jupiter and Redstone hardware (propellant tanks) using eight H-1 rocket engines derived from the Jupiter missile. The massive first stages were static tested at Huntsville and shipped by barge to Cape Canaveral for launch. The first four 162 foot tall vehicles were launched with dummy upper stages from Complex 34 in 1961-63. Total thrust was 1,330,000 pounds, the greatest thrust for an American rocket at the time. Six more Block II Saturn I's (190 ft tall) were flown from Complex 37 with Douglas S-IV upper stages propelled by five RL-10 hydrogen-powered engines from the Centaur program. Thrust generated was 90,000 pounds. The uprated Saturn IB version featured the S-IVB upper stage (intended for the Saturn V) and flew nine missions before they were retired by NASA in 1975. Notable achievements included the first orbital flight test of the Apollo Lunar Module (Apollo 5) and the first manned test flight (Apollo 7) in October 1968.

First Saturn (SA-1) is shown inside the gantry at launch Complex 34 prior to the successful first flight with a live first stage and dummy upper stages on October 27, 1961.

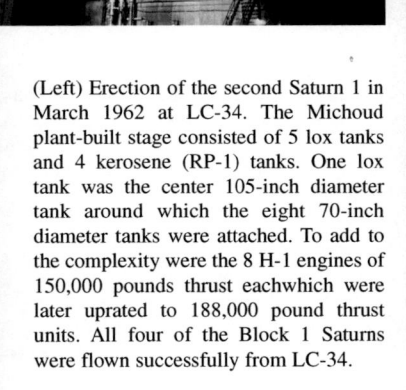

(Left) Erection of the second Saturn 1 in March 1962 at LC-34. The Michoud plant-built stage consisted of 5 lox tanks and 4 kerosene (RP-1) tanks. One lox tank was the center 105-inch diameter tank around which the eight 70-inch diameter tanks were attached. To add to the complexity were the 8 H-1 engines of 150,000 pounds thrust eachwhich were later uprated to 188,000 pound thrust units. All four of the Block 1 Saturns were flown successfully from LC-34.

(Right) Preparing to hoist the first Douglas built S-IV stage onto SA-5 in late 1963 at LC-37B. The five engine RL-10 cluster was very similar the Centaur RL-10 engines (flown for the first time in November 1963). This stage and the remaining 5 flight stages were all successful on the six block II Saturns flown as SA-5 through SA-10 in 1964-1965.

(Above) Detailed view of Saturn I vehicle SA-2 at pad 34 in March 1962, with the tail service mast in place during a pre-launch test. The remarkable open-air service structure looms in the background. Note the details of the dummy upper stages.

(Right) Saturn SA-8 lifts off with a Pegasus payload concealed beneath a dummy Apollo spacecraft.

(Left) With its oversized fins the first two stage Saturn poses for its photograph prior to flight January 29, 1964. SA-5 was a complete success and carried a dummy payload of sand ballast into orbit. Despite its sterling record the S-IV second stage was never used beyond the six flights on Block II Saturns.

(Below) The first and only night launch of a Saturn I/IB was SA-8 at 2:35 on May 25, 1965 from LC-37B. Here a time exposure from the north shows the new LC-39 complex still under construction. Visible is the Vertical Assembly Building and the three Saturn V launch umbilical towers. The payload was a Pegasus micrometeroid research satellite. SA-8 was actually the ninth Saturn flown as SA-9 had preceded it.

(Left) Carrying the first boilerplate Apollo spacecraft for aerodynamic evaluation, Saturn SA-6 is captured by a remote camera atop the pad 37B umbilical tower during liftoff on May 28, 1964. The blockhouse is visible in the upper right hand corner.

(Right) Saturn IB was an uprated version of Saturn I that was intended to carry astronauts during early Apollo test flights. It utilized the powerful liquid hydrogen/liquid oxygen J-2 engine in the new and larger S-IVB upper stage built by Douglas, that would serve as the third stage of the Saturn V beginning in 1967. This particular vehicle is AS-203, bathed in the floodlights of pad 34 in August 1966.

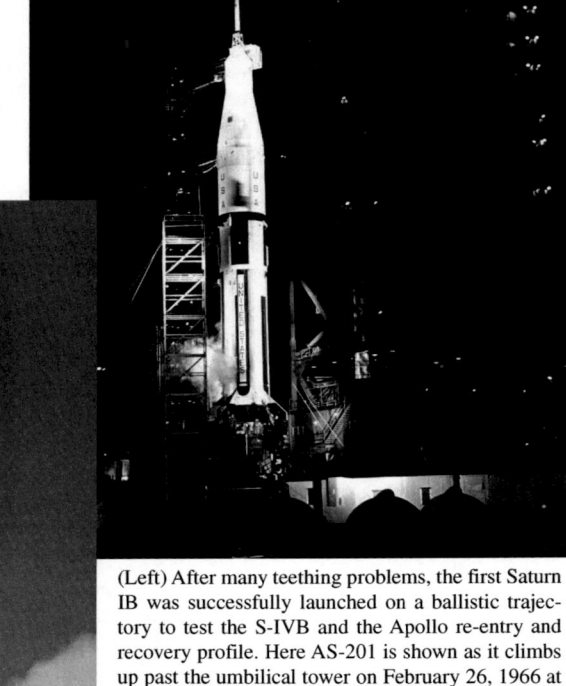

(Left) After many teething problems, the first Saturn IB was successfully launched on a ballistic trajectory to test the S-IVB and the Apollo re-entry and recovery profile. Here AS-201 is shown as it climbs up past the umbilical tower on February 26, 1966 at 11:12 AM. The flight lasted 32 minutes from lift off until splashdown.

(Right) A view looking up at the cluster of eight H-1 engines on the first stage of the Apollo 5 launch vehicle (AS-204) from beneath the pad 34 launch pedestal at Cape Canaveral. The vehicle is suspended over the mounting ring by four hydraulic hold-down arms.

(Left) The second Saturn IB to fly was AS-203 on a test of the S-IVB upper stage. The objective was to observe the behavior of the propellants in zero gravity during the parking orbit between burns as would occur on an Apollo flight to the moon. The July 5, 1966 launch was completely successful.

(Right) On January 22, 1968 the Apollo lunar module underwent its first test flight in low-earth orbit from LC-37 with the Saturn AS-204 booster, originally allocated to the Apollo 1 flight. The Grumman-built lunar lander was subjected to a number of maneuvers to prove out the design. The Apollo 5 flight was a success, paving the way for a manned test of the LM on the third Apollo test flight in March 1969.

(Below) Apollo 7 clears the tower on October 11, 1968 in this close up launch sequence from the top of the umbilical tower. The mobile service tower is visible in the background.

(Right) Apollo 7 ascends on a pillar of fire as the Saturn IB rocket lifts off from Complex 34 to begin the first manned flight of the Apollo program on October 11, 1968 (test 066). Astronauts Schirra, Eisle and Cunningham completed a very successful 11 day flight to set the stage for Apollo 8's triumphant lunar voyage in December 1968.

(Below) A remarkable long-range aerial portrait of the Apollo 7 launch captured by an Air Force ALOTS tracking aircraft flying at 35,000 feet altitude south of pad 34. The closeness of the Vertical Assembly Building in the background at KSC is an illusion created by the powerful optics of the ALOTS tracking telescope.

THOR "BIG SHOT"
"Big Shot" was two suborbital test flights of the enlarged 135 foot Echo balloon design, flown by NASA on Thor rockets in January and July 1962. Formally designated the Applications Vertical Tests (AVT), the first flight failed when the balloon membrane ruptured, but the second flight was successful and qualified the design for orbital flight. Echo 2 was launched from Vandenberg AFB in January 1964. Television cameras monitored each balloon as it inflated, and a capsule with a motion picture camera was deployed to film each test. The capsules were recovered from the Atlantic by Air Force air-sea rescue men.

(Left) Technicians fit the fairing around the first Big Shot balloon canister in one of the Douglas hangars at Cape Canaveral in November 1961.

(Right) The single-stage Thor (number 337) for the first Big Shot mission is prepared for installation at pad 17A in January 1962.

(Left) Spectacular nighttime liftoff for the second Big Shot balloon test on July 18, 1962 from pad 17A (test 082). The Echo balloon was successfully deployed after the first balloon burst during the inflation process.

(Right) Motion picture still of the ruptured balloon during the first Big Shot test in January 1962.

THOR-ASSET
ASSET (Aerothermodynamic/elastic Structural Systems Environmental Tests) was a series of six Air Force reentry research flights conducted with one and two stage Thor rockets (returned from overseas deployment with the Royal Air Force in Britain) at Launch Complex 17B from September 1963 to May 1965. ASSET was a sub-scale lifting body vehicle (without wings) that was used to test thermal protection methods for future spacecraft designs. It was originally designed to support the Dyna-Soar (X-20) program. Three ASSET test flew on single-stage Thor boosters, and three tests utilized two-stage Thor-Delta vehicles to provide additional velocity during entry. Four of the six shots were successful (one was destroyed by range safety) but only one of the three scheduled recoveries was actually accomplished.

(Right) An ASSET test vehicle is examined by a technician at the top of the gantry at pad 17B.

(Left) The first Thor-ASSET vehicle (ASV-1) was erected on pad 17B in late August 1963 in preparation for launch on September 18, 1963.

(Below) McDonnell technicians manhandle the second ASSET glider from the checkout building on the way to the launch pad. Note the sporty ASSET logo displayed in several locations in this picture.

(Left) An ASSET glider and its payload adapter arrive at the top of the gantry at pad 17B prior to mating with the launch vehicle.

(Right) Liftoff of a single-stage Thor rocket at pad 17B with the fifth ASSET glider in the series on December 8, 1964.

(Below) Payload adapter of the second ASSET vehicle (out of frame) is mated to the Delta upper stage at pad 17B. The first Thor-Delta ASSET flight failed during staging on March 24, 1964 (test 075).

(Right) The final ASSET vehicle (ASV-4) sits atop the Thor-Delta rocket at pad 17B in February 1965 (test 0175).

(Left) Two-stage Thor-Delta launch with an ASSET payload from pad 17B.

PROJECT FIRE (ATLAS-ANTARES)
Project FIRE (Flight Investigation of the Reentry Environment) was the first reentry experiment to reach 25,000 mph, the velocity of a spacecraft returning from the moon. NASA conducted the two FIRE experiments on April 14, 1964 and May 2, 1965 using Atlas D boosters with Antares II (X-259) upper stages. After a 20 minute coast following booster sep, the Antares motor was fired in a downwards direction to accelerate the FIRE payload to a nominal velocity of 31,000 feet per second, approximately 25,000 mph, in the vicinity of the Ascension Island tracking station. Sophisticated optical sensors recorded the fireball of each FIRE vehicle to gather data about high speed reentry phenomena, and to verify the underlying principles used to design the Apollo heat shield. The experiments were successful in both respects. Each 190 pound FIRE vehicle, built by Republic Aviation, was 26 inches in diameter and 21 inches long, shaped like a miniature Apollo command module. No recovery was planned for either vehicle.

High angle view of Atlas 264D from the service tower at pad 12 prior to the second FIRE experiment in May 1965.

(Right) The first FIRE reentry vehicle is mated to the Antares II solid rocket kick motor that would be used to accelerate the FIRE vehicle to the equivalent velocity of a return from the moon.

(Above) The second FIRE vehicle at pad 12, before it was mated to the velocity package and hoisted to the top of the gantry.

(Right) The velocity package and Project FIRE projectile are hoisted to the pad 12 white room to be mated with Atlas 233D in April 1964.

(Left) Activity in the pad 12 blockhouse during a practice countdown for the FIRE 1 mission in late March 1964.

(Below) Liftoff of the FIRE 1 reentry experiment at 4:42 pm EST on April 14, 1964 from pad 12 (test 0225).

(Below) Project FIRE's second launch on May 22, 1965 at 4:55 pm, with Atlas 264D (test 0501). Both missions were successful.

POLARIS-STAFF

Project STAFF (Stellar Acquisition Flight Feasibility) was an unusual situation in that it was an Air Force project that hitched a ride on a Navy Polaris A-1 missile at Cape Canaveral (then called Cape Kennedy). The purpose of the collaboration was to test the Stellar Inertial Guidance System (STINGS) on four Polaris flights from pad 29A in 1965. The first flight on April 14, 1965 was destroyed by range safety at 73 seconds, but the next three flights were successful. The star tracker mechanism locked on to the primary target star (appropriately, Polaris) and provided data that was used to refine ground based trajectory data. STINGS was developed for mobile ballistic missile systems, but these weapons were never developed and the potential of stellar inertial guidance was never realized.

(Left) The third STAFF launch (STP-3) from pad 29A on August 11, 1965. The Polaris A-1 launch vehicles (designated C2X) used for Project STAFF had previously served as live nuclear missiles during operational Polaris submarine cruises.

Both photographs below depict the second STAFF payload (STS-2) launched on June 16, 1965 from pad 29A. The STINGS sensor head has yet to be covered by the conical payload fairing. The pictures were taken from inside the small service structure at Complex 29.

ATLAS-AGENA AT PAD 13

After the last Atlas E R&D missile (40E) flew from Launch Complex 13 on Feb, 13,1962, the Air Force demolished the original service tower to replace it with a new structure to support three Atlas-Agena launches for the Defense Department. After the rebuilding project was completed in February 1961 the first Vela nuclear detection satellite was launched from LC-13 in October 1961. NASA began operating from pad 13 in November 1964 with the launch of Mariner 3 (an unsuccessful Mars probe). After five Lunar Orbiter spacecraft and an Orbiting Geophysical Observatory, NASA relinquished control of Complex 13 back to the Air Force in 1968 for a series of classified payloads from August 1968 to April 1978. Complex 13 remained the only intact, original Atlas facility until the umbilical tower was demolished due to safety and environmental concerns in December 2004, followed by the service tower in August 2005.

(Right) The new mobile service tower at Complex 13 (modeled after the tower at Complex 36) while under construction in late 1962.

(Left) The first Vela nuclear detection payload atop an Atlas-Agena rocket as the 179 foot service tower is retracted from pad 13 in October 1963.

(Right) The second Atlas-Agena rocket at pad 13 at the beginning of the LOX loading procedure. The twin vela payloads were stowed under an unusual reddish-colored fiberglass fairing. Liftoff of the second Vela mission took place on July 17, 1964 (test 2925).

(Below) Liftoff of Vela payload at night from pad 13.

(Right) Polished metal fairing on a classified Atlas-Agena launch from pad 13 on May 23, 1977 (test 0694).

(Left) Distinctive extra-long fairing on several of the classified Atlas-Agena's launched from pad 13 are believed by analysts to conceal an enormous furled mesh eaves-dropping antenna installed on a geosynchronous spacecraft code-named Rhyolite.

TITAN 34D & COMMERCIAL TITAN

Titan 34D was the Air Force's replacement for the highly effective Titan IIIC heavy lift booster that was retired in 1982. Martin Marietta achieved a significant gain in payload capacity by lengthening the basic Titan core stage by six feet and developing new 5½ segment solid rocket motors with the manufacturer, United Technologies. The extra half-segment increased the thrust of each motor from 1.2 to 1.4 million pounds with the same 116 second burn time. Titan 34D was in service for nearly seven years at Cape Canaveral, flying a total of 16 missions. Titan 34D was eventually replaced by the Titan IV in 1989, but Titan 34D received a new lease on life in August 1987 when Martin Marietta formed the Commercial Titan launch services company to try and capture a portion of the world's launch-for-hire market. The first Commercial Titan vehicle based on Titan 34D flew on December 31, 1989 on a successful mission to deploy two communication satellites, but the second mission on March 14, 1990 stranded the Intelsat 603 payload in the wrong orbit due to incorrect interface wiring in the release mechanism. This is the satellite that the space shuttle astronauts rescued and reboosted to the proper orbit in May 1992. Commercial Titan managed to launch the Intelsat 604 payload successfully in June 1990, but the only other customer the company could sign up was the ill-fated Mars Observer space probe for NASA that was launched in September 1992 (see next page).

(Left) The first Titan 34D booster with the new 5½ segment solid motors lifts off from pad 40 on October 30, 1982 on a completely successful maiden flight with two military communication satellites.

(Right) NASA's Mars Observer launch from pad 40 on September 25, 1992 was the fourth and final mission for the Commercial Titan program.

(Above and lower right) The first Titan 34D flight marked the introduction of the massive Boeing solid motor Interim Upper Stage (IUS), seen here during the hoist up to the white room of pad 40 to be mated with the Titan core stages. The VIB and SMAB processing buildings for Titan can be seen on the horizon in the upper left hand picture.

(Left) The second stage of the commercial Titan is being lifted into position inside the Vertical Integration Building.

(Above) Aerojet's Transtage upper stage on the transport trailer in the Vertical Integration Building. The bi-propellant rocket was the primary upper stage for Titan IIIC and Titan 34D until it was replaced by the IUS stage on the follow-on Titan IV launch vehicle in the late 1980's.

(Left) The Commercial Titan vehicle for NASA's ill-fated Mars Observer spacecraft is enclosed within the Mobile Service Structure in this unusual portrait at pad 40.

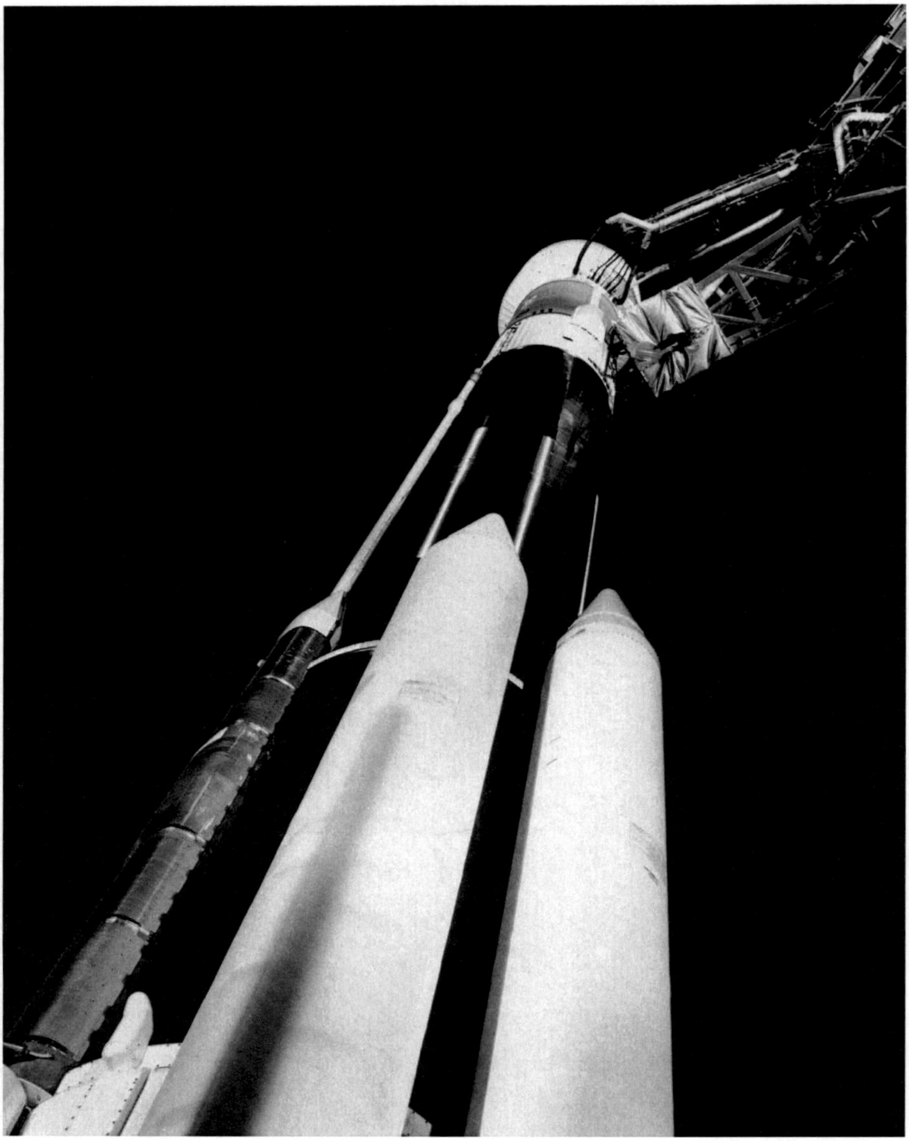

Part III

SPACE LAUNCH IN THE POST CHALLENGER ERA (1989-Present)

TITAN IV

Titan IV was the third major variant of the Titan IIIC rocket to fly from pads 40 and 41 at Cape Canaveral since 1965. The Complementary Expendable Launch Vehicle (CELV) was instigated by Secretary of the Air Force Pete Aldridge in 1985 when it became clear to Pentagon space officials that the Space Shuttle was not a reliable launch vehicle for national security payloads (the Shuttle could not lift heavyweight payloads into polar orbit). The decision to build the CELV, which was named Titan IV by Martin Marietta (later Lockheed Martin), was vindicated when the Challenger exploded shortly after liftoff in January 1986. The Titan IVA featured larger 7 segment solid booster motors and stretched Titan core stages, while Titan IVB (introduced in 1997) had larger (10.5 foot diameter), more powerful Solid Rocket Motor Upgrade (SRMU) boosters provided by Alliant Tech Systems. Although intended for military payloads, Titan IVB also launched the Cassini probe to Saturn in October 1997. Titan IV was flown with either the Interim Upper Stage (IUS) or Centaur G upper stage (and on rare occasions with no upper stage for certain classified payloads). The final Titan IV was launched from pad 40 at Cape Canaveral on April 29, 2005. Titan IV has been replaced by a combination of Delta IV and Atlas V boosters.

(Left) Titan IVB booster arrives at pad 40 without the Centaur upper stage, which was added later within the confines of the Mobile Service Structure.

(Right) Spectacular night-time departure of the Cassini space probe to Saturn from pad 40 on October 15, 1997. The vehicle was designated K-33.

(Left) Titan IV preparations began with attachment of the SRMU boosters in the Solid Rocket Motor Assembly and Readiness Facility (SMARF).

(Below right) Twin locomotives pull the launch platform and Titan IV vehicle to pad 40 where launch processing would continue.

(Below left) The Centaur G upper stage arrives at the base of pad 40 prior to being hoisted to the top of the service structure for mating to the Titan core stages.

(Right) The 16.7 foot diameter payload fairing is lowered into position over the Cassini space probe within the spacious interior of the pad 40 Mobile Service Structure.

(Above) Looming like a twenty story office building at pad 41, this Titan IVA rocket dwarfs the umbilical mast attached to the transporter platform.

(Right) A camera at the top of the umbilical mast captures the moment of liftoff for the first Titan IVB on February 23, 1997. Smoke and fire belch from the flame duct far below as the classified payload successfully launched into orbit from pad 40.

(Left) The first Titan IVB rocket with its massive 112 foot high SRMU boosters totally dominate the facilities at pad 40. Each SRMU booster was 10.5 feet in diameter.

(Right) Developing 3.4 million pounds of thrust at liftoff, Titan IV was the most powerful expendable launch vehicle ever built (Saturn V and the N-1 were built for manned spacecraft). The photograph depicts the first Titan IVB vehicle (B-24) towering over the ground facilities at pad 40. One of the 350 foot lightning protection towers is visible at right.

DELTA II OPERATIONS AT PAD 17
Delta II was developed as a direct consequence of the Challenger shuttle disaster. President Ronald Reagan decreed that commercial space payloads would no longer be permitted to fly on the shuttle, and NASA was directed by Congress to re-open the production lines of the Delta and Atlas-Centaur expendable launch vehicles. The first Delta II flew in February 1989 from Complex 17, and is still operational today, the only so-called "heritage" vehicle still flying in the 21st century (Delta II is a direct descendant of the Thor intermediate range missile). Delta II can fly with up to nine solid rocket motors and a variety of payload fairings.

(Right) An early version of the Delta II (number 210) at pad 17A in June 1992.

Delta II stacking operations: the first stage of the Delta II is brought to Complex 17 (left) and is suspended in the mobile service structure (lower right) prior to being moved into position over the launch mount at the pad for final mating (opposite top).

Mounting the solid rocket motors to the Delta II first stage: each individual motor is also suspended inside the service structure (bottom right) and the entire gantry is moved up to the rocket where the motors are attached (bottom left). The photo below illustrates hoisting of the payload (shiny cylinder) up the side of the gantry by a small bridge crane. Once at the top, the crane is moved over to the center section and lowers the payload onto the rocket. The Delta II upper stage is also installed in the same way.

Two views of the business end of a Delta II rocket at Complex 17. In the top photo one of the large solid rocket motors is slowly being moved to mate with the mounting fixture (socket) on the side of the vehicle. Below is a close up of the solid rocket nozzles and the Delta II main engine on the deck of the pad at Complex 17.

Spectacular liftoff of the Swift gamma-ray astronomy satellite in November 2004 from pad 17A, on a Delta II model 7320 vehicle with three SRM's.

ATLAS I, II, III

In 1986 President Ronald Reagan reinstated all expendable launch vehicle production in the wake of the shuttle Challenger disaster when rocket production lines were being closed in favor of the Space Shuttle. After Delta reappeared in 1988 and Titan IV in 1989, Atlas commercial space launches by General Dynamics commenced in 1990. The re-named Atlas I vehicle was a continuation of the Atlas G - Centaur that existed before the shuttle accident. In short order General Dynamics introduced the Atlas II version with elongated propellant tanks and an improved Centaur stage, followed by the Atlas IIA model with further improvements. In December 1993 General Dynamics introduced its most powerful Atlas variant, the Atlas IIAS, with three solid rocket motors that improved the Atlas payload capacity by 1500 pounds. Atlas IIAS later became the standard Atlas version until the introduction of the Atlas III in 2000, flying a total of 30 missions until it was retired by Lockheed Martin in August 2004. A total of 77 vehicles flew from LC-36 in the commercial launch program. Atlas III was a radical departure from the usual American rocket designs incorporating Russian RD-180 rocket engines attached to the classic steel balloon tank Atlas structure. The Atlas III design was intended by Lockheed Martin (a merger between the manufacturers of Titan and Atlas rockets) to "retire the risk" of installing Russian engines in the new Lockheed Martin Evolved Expendable Launch Vehicle (EELV) that later became the Atlas V. Atlas III was wholly successful, flying the last of its six missions in February 2005, which also retired the original "heritage" version of Atlas after nearly 48 years in service.

Atlas commercial launch activity was inaugurated on July 25, 1990 at Cape Canaveral with the successful liftoff of AC-69 carrying the joint Air Force - NASA CRRES research payload. Note the new 12 foot diameter payload fairing introduced for the Atlas I series.

(Top) Technicians inspect the engine compartment of a latter-day Atlas I vehicle after it arrived at Complex 36.

(Left) An Atlas I rocket is slowly winched into a vertical position on the launch pad, protected within the service structure at Complex 36.

(Right) A Centaur stage arrives at Complex 36 for mating with the Atlas I first stage.

(Left) An older version of the Centaur stage without its insulation panels is hauled up to the top of the gantry by crane for mating with the launch vehicle.

(Above) Beginning with the Atlas II version, the Centaur stage flew with spray-on foam insulation in place of the original jettisonable insulation panels.

(Left) Atlas II was an upgraded version of the standard Atlas I vehicle with elongated first stage propellant tanks. This is vehicle AC-103.

(Right) One of the later Atlas II vehicles, AC-116, soars aloft from pad 36A on May 31, 1995 carrying the Navy's EHF Follow-on communications satellite (Op. No. A2052)

(Above) The slender profile of the Atlas IIA booster is revealed at pad 36B as the mobile service structure is rolled back prior to launch on the afternoon of September 8, 1993.

(Right) Atlas IIA vehicle AC-120 carries the Galaxy III-R commercial satcom payload towards orbit on December 14, 1995.

AC-108 was the first Atlas to fly with solid rocket motors, which lifted off from pad 36B on December 15, 1993. The three Castor 4A solid motors increased the Atlas IIAS payload capacity by 1500 pounds.

(Above) The modernized interior of the Complex 36 blockhouse was utilized for Atlas Centaur launches from 1990 to 2005. The gentleman in the foreground is Skip Mackey, who read telemetry reports live on the NASA launch commentary for many years until his recent retirement.

(Left) AC-206 enacts the final chapter in the original Atlas story as it lifts off in the fog on the final Atlas mission from Complex 36B on February 3, 2005 with a classified payload from the National Reconnaissance Office.

DELTA III

Delta III was a failed attempt by Boeing to offer an uprated version of their reliable Delta II rocket in the late 1990's, with a cryogenic upper stage and enlarged solid rocket motors. When the first two launch attempts failed with the loss of their commercial payloads, Boeing arranged a test flight from pad 17B in August 2000 to reassure potential customers. When the dummy demonstration payload was placed in a simulated geosynchronous transfer orbit that was at least 10,000 miles too low, demand for the rocket evaporated and the program was eventually terminated by Boeing. The remaining Delta III rockets were cannibalized for parts and the large solid rocket motors were bequeathed to the Delta II Heavy variant. Extensive modifications were required at pad 17B to support the 128 foot launch vehicle: new propellant fill lines to service the cryogenic second stage, increased water deluge flow and enlargement of the solid motor openings in the launch platform.

The third Delta III launch vehicle (Delta 280) stands tall at pad 17B on the eve of its demonstration test launch on August 23, 2000. With twice the payload capacity of the current Delta II model, Delta III would have been a useful addition to the U.S. launch vehicle fleet, but it failed to attract any new customers after two launch failures, and Boeing quietly pulled the plug on their expensive investment.

ARIES, STARBIRD AND PROSPECTOR AT LC-20

Launch Complex 20 was a former Titan I and Titan IIIA site that was selected by the Air Force in 1988 to host sounding rocket launches for the Starlab military research mission on the Space Shuttle. The first of two 58 foot heavy launch rails for the four-stage Starbird rocket was installed at LC-20 in September 1990 in preparation for a test firing that December. The flight was successful, but the Starlab program was cancelled by the Pentagon and the Starbird rockets evidently never flew again. In June 1991 a commercial microgravity research payload called Joust was destroyed after only 25 seconds of flight on a Prospector rocket (based on the Castor solid rocket motor). Additional Joust flights were cancelled when the program was eliminated from the NASA budget. Two Aries solid rockets were launched from LC-20 in August and October 1991 for the classified Red Tigress program for the Air Force. The final two launches from complex 20 were made in May 1993 for Red Tigress II using four-stage Low Cost Launch Vehicles (LCLV). The site is presently inactive, but is available for lease from the Florida Space Authority (a state agency).

(Left) Aries rocket at Complex 20 for the Red Tigress program in 1991.

(Below) The Starbird development vehicle lifts off from Complex 20 on December 17, 1990 (test 6406).

ATHENA AT LC-46
Lockheed Martin's Athena rocket was only the second solid fuelled orbital launch vehicle to fly at Cape Canaveral (the first was a single unsuccessful Blue Scout I in November 1961). Athena 1 was a 62 foot two-stage rocket and Athena 2 had three stages with a height of 93.2 feet. Athena flew only twice at Launch Complex 46 before the market dried up for small satellite launches. The first launch was NASA's Lunar Prospector in January 1998 on an Athena 2 version, while the second was for the Republic of Taiwan. The launch site was established by the Florida Space Authority (FSA) at the former location of Navy Trident II testing (1981-89). LC-46 is currently maintained in caretaker status by the FSA in case the small satellite market revives in the near future.

(Right) The Athena 2 launch vehicle for NASA's Lunar Prospector mission stands ready at Complex 46 with the newly constructed service structure in the background.

(Left) Liftoff of Lunar Prospector at 9:28 pm on January 6, 1998.

ATLAS V

Atlas V is Lockheed Martin's replacement for the retired Titan IV heavy-lift booster. In conjunction with Boeing's Delta IV, Atlas V represents a new generation of American rockets known as the Evolved Expendable Launch Vehicles (EELV), developed in response to an initiative by the Pentagon in 1998 to replace the existing Atlas and Titan launch vehicles with more efficient and economical rockets. Lockheed Martin built their vehicle around the Russian RD-180 rocket engine that had already been proven in the Atlas III series. The engine develops 860,000 pounds of thrust in the baseline first stage called the Common Core Booster (CCB). Atlas V retained the high-energy single engine Centaur II upper stage from Atlas III. The common core is 12.5 feet in diameter (compared with 10 feet for Atlas III) and can accommodate up to five Aerojet solid rocket motors. The Atlas V model 551 with 5 SRM's develops 2.1 million pounds of thrust at liftoff and is the most powerful Atlas V variant. Lockheed Martin borrowed the Russian concept of the "clean pad" to process the Atlas V for flight. There are minimal fixtures at the launch pad beyond the launch mount and exhaust duct - all vehicle processing is conducted in the Vertical Integration Facility (VIF) located 1,800 feet from launch pad 41, the former Titan IV facility. The vehicle is rolled to the pad typically less than 24 hours before the scheduled launch. The first Atlas V flew in August 2002.

Liftoff of the Mars Reconnaissance Orbiter (MRO) from Launch Complex 41 on August 12, 2005. MRO was NASA's first use of the new-generation Atlas V launch vehicle.

(Below) Lockheed Martin adopted a slightly different approach for processing their new Atlas V vehicles at Canaveral, taking their inspiration from the Russians who checkout their launch vehicles in the assembly building and only transport them out to the launch pad at the last possible moment. Here the Atlas V core stage has arrived at the Vertical Integration Facility (VIF) where it is erected on the launch table (on a transporter) and mated with the Centaur upper stage and payload.

(Below right) The Atlas V core stage has been lifted from the transport vehicle at the VIF prior to installation on the launch table. The vehicle in these photos is AV-007 of the Mars Reconnaissance Orbiter.

(Right) The single-engine Centaur upper stage has arrived at the VIF and is in the process of being lifted into the VIF to be mated with the Atlas V core stage.

(Above) The Atlas V launch Vehicle for MRO has just arrived at pad 41 after travelling the 1,800 feet to the launch pad from the VIF on rails less than 12 hours prior to launch. The umbilical mast is attached to the transport platform.

(Right) The Pluto New Horizons spacecraft, encapsulated within the payload fairing, is gently lowered to the Atlas V adapter inside the VIF structure at Complex 41 in late 2005.

(Upper left) A 67 foot Aerojet solid rocket motor for Atlas V is hoisted into position inside the Vertical Integration Facility (VIF) to be attached to the launch vehicle of Pluto New Horizons (AV-010) in November 2005.

(Center left and below left) Rollout of the Atlas V vehicle for Pluto New Horizons on the day before scheduled liftoff in January 2006.

(Right) The Atlas V Spaceflight Operations Center at Cape Canaveral accommodates the launch controllers in safety and comfort (a blockhouse is no longer required right at the pad to control the launch).

(Left) The Atlas V rocket for the Pluto New Horizons mission (AV-010) is ready for launch at pad 41 one day prior to the scheduled launch on January 17, 2006 at Cape Canaveral (which was delayed two days to January 19).

The spectacular liftoff of the Pluto New Horizons probe (aided by five strap-on solid motors) from pad 41 on January 19, 2006 at 2:00 pm local time.

DELTA IV

Delta IV is Boeing's entrant in the Evolved Expendable Launch Vehicle (EELV) program, and competitor to Lockheed Martin's Atlas V rocket. The Common Booster Core (CBC) of Delta IV is 15 feet in diameter and is powered by a single Rocketdyne RS-68 hydrogen-oxygen rocket engine, the first large American liquid propellant rocket engine developed since the space shuttle main engines in the 1970's. Each engine develops 657,650 pounds of thrust at liftoff, less powerful but more efficient than the Atlas V powerplant. Up to four Alliant GEM 60 solid rocket motors can be attached to the CBC stage. The hydrogen-oxygen upper stage was adapted from Delta III for use with Delta IV, powered by the RL-10B2 engine of Centaur fame. The core stages are designed to be modular, and the Delta IV Heavy version is comprised of three CBC stages attached together side-by-side. The Delta IV Heavy vehicle develops 1,973,000 pounds of thrust at launch. The maiden flight in December 2004 was marred by sloshing in the propellant tanks that reduced engine performance significantly. Delta IV varies in several other aspects from Atlas V. Delta IV's propellant tanks are insulated by spray-on foam insulation much like the space shuttle external tank uses (Atlas V has a conventional aluminum structure). Atlas V is delivered by aircraft from the Lockheed Martin factory near Denver, while Delta IV is delivered to the Cape by ship from the factory at Decatur, Alabama. Launch facilities for Delta IV are more traditional than for Atlas V, with vehicle processing conducted on the launch pad much like Delta II, only on a larger scale. The first Delta IV vehicle flew from Complex 37 in November 2002.

(Left) Stately liftoff of a Delta IV Medium launch vehicle between the lightning towers at pad 37B.

(Right) The fiery and dramatic maiden launch of the Delta IV Heavy vehicle 1. on December 21, 2004. The charred appearance of the triple-mounted Common Core Boosters is due to hydrogen vapors igniting the spray-on insulation coating the boosters (perfectly normal, according to Lockheed Martin engineers). The flight was marred by excessive sloshing of the propellants in the tandem stages, and engineers have scrambled to devise a solution to the potentially catastrophic problem.

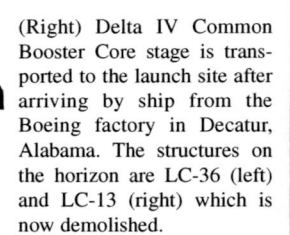

(Right) Delta IV Common Booster Core stage is transported to the launch site after arriving by ship from the Boeing factory in Decatur, Alabama. The structures on the horizon are LC-36 (left) and LC-13 (right) which is now demolished.

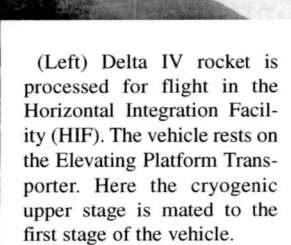

(Left) Delta IV rocket is processed for flight in the Horizontal Integration Facility (HIF). The vehicle rests on the Elevating Platform Transporter. Here the cryogenic upper stage is mated to the first stage of the vehicle.

Common booster core inside the Horizontal Integration Facility near Complex 37.

The Horizontal Integration Building at the new Delta IV installation at Complex 37. The Pathfinder version of the Common Booster Core is parked outside the HIF at the present time.

(Right) A complete Delta IV rocket pulls out of the HIF on its KAMAG road transporter for the short journey to pad 37B on Beach Road at Cape Canaveral.

(Left) The Delta IV vehicle is raised to firing position at pad 37B by the hydraulic platform of the road transporter. The encapsulated payload is later delivered to the pad for installation on the launch vehicle.

PRESS COVERAGE AT THE CAPE

The news media were not permitted on-base to cover launches at Cape Canaveral (on a regular basis) until March 1958. Prior to that date observers had to surreptitiously photograph launches with long lenses from the nearby beaches: hence their famous nickname "the birdwatchers." Three designated press sites were eventually built to accommodate the news media at Canaveral, including network television, covering the Mercury and Gemini manned flights and unmanned satellite launches. The first televised launches were the Pioneer moon shots in August and October 1958 on a tape-delay basis. Most launches today can be viewed on satellite broadcasts or internet streaming video provided directly from the rocket manufacturers or NASA. In the 1990's even previously classified space launches on Titan or Atlas vehicles were allowed to be covered on television and witnessed by reporters, in a total reversal of Air Force public affairs policy dating from the late 1950's. Presently the press viewing areas are located near LC-17 (for Delta II), on the NASA Causeway across the Banana River (for Delta IV) and the LC-39 press site at Kennedy Space Center for Atlas V.

(Left) Long lenses of the "birdwatchers" peering toward an early Atlas rocket from the local Cape Canaveral beaches in 1957.

(Right) News media cameramen at Press Site 1 strain to follow the swiftly accelerating Delta II rocket bearing the MAP astrophysics probe on the afternoon of June 30, 2001.

(Left) Press Site No. 2 at Cape Canaveral several days before the launch of Gordon Cooper on Mercury-Atlas 9. The three launch towers on the horizon are for Titan II. LC-14 for Mercury-Atlas is just out of the picture on the right.

(Right) Reporters gathered at Press Site 2 for a Gemini launch in 1965. Note the CBS color TV camera on the raised platform.

The view from the press site: (left) Delta 221 accelerates at a terrific pace from pad 17 carrying a Navstar GPS payload to orbit on June 26, 1993. The author used a 1000 mm lens at Press Site 1 to capture the Delta II in flight just seconds after it cleared the tower.

(Right) An Atlas I vehicle during launch is silhouetted against the Florida sky at sunrise on September 3, 1993 as also seen from Press Site 1. AC-75 was launched from pad 36B. The bird on the wire had a better vantage point than the news media did.

(Left) Launch Complex 17 is the closest facility to Press Site 1 and provides the best launch viewing at Cape Canaveral (except perhaps the Delta IV viewing site on the NASA Causeway). The photo shows the night launch of Delta 273 with a Globalstar payload from pad 17A in July 1999.

The Ruins of Cape Canaveral

(Right Top) This is one of the most historic spots at Cape Canaveral, the location of the first launches in July 1950 (Bumper rockets, based on the German V-2). The site, retroactively designated Complex 3 after the long Range Proving Ground was established, is located near the famous Cape lighthouse, just south of the concentration of Atlas, Titan and Saturn pads known as "ICBM Row."

(Left) A Thor-Delta mockup resides today at pad 26A, site of America's first Earth satellite launch, Explorer 1, on January 31, 1958. The Air Force Missile and Space Museum now occupies the entire area comprising Complex 26 and the Army's Complex 5/6 (site of the first two manned Mercury Redstone suborbital launches in 1961).

(Below) With the present day gantry of pad 17A (Delta II) looming in the background, the remains of the Vanguard launch site (pad 18A) are visible in these two photographs. The concrete outbuilding (24403, bottom) is still used by the Air Force today for storage. The fading striped pylon of the water deluge system is prominent in many contemporary photographs of the Vanguard launch vehicle.

(Right) Abandoned fuel storage revetments and other concrete fixtures are all that remain of the Thor tactical launch site (pad 18B) that saw its last launch in January 1960. The Blue Scout I and II suborbital research rockets were also launched at pad 18B in the early 1960' s.

(Left) The imposing concrete hard stand (pad 9) of the Navaho missile, stripped of all useable equipment save for the stairway handrails, is one of the most prominent "ruins" at Cape Canaveral today. The central cavity is the flame channel for the missile's exhaust.

Blockhouses: The forward portion of the Navaho blockhouse facing the launch pad (Complex 9). The extended sections held mirrors to reflect the scene to the launch controllers inside the blockhouse.

(Left) Looking vaguely Mayan, the outer shell of the pad 34 blockhouse as it appears in the present day. The launch complex was closed in 1971 after the Skylab launches were assigned to LC-39B at Kennedy Space Center.

(Right) Curved outlines of the twin rails for pad 10's portable Navaho work platform (outlined in grass) are still visible today at that location, which is also the site of the deactivated Minuteman complex (silos 31 and 32) that was built there in 1959, obliterating the former Navaho tactical launch site.

The remains of Minuteman's Complex 32. The concrete structure at ground level is the cap placed over the silo opening after the wreckage of Challenger was entombed there by NASA in 1987. The central "beehive" structure is the former pad 32 blockhouse, and the rectangular structure in the background is pad 9 of Navaho. The Navaho blockhouse is at right.

(Right) Workers prepare the old Minuteman silo at Complex 32 to receive the remains of the space shuttle Challenger in early 1987.

(Left) A piece of the fuselage side wall of Challenger is lowered into the former Minuteman silo at Cape Canaveral.

Little remains of the historic Complex 14 site except the framework of the launch pad and the ramp leading up to the pad structure. Mercury astronaut John Glenn and three of his colleagues were launched into orbit atop Atlas missiles from this site in 1962-63.

(Left) The mobile service structure (gantry) at pad 14 succumbed to salt air corrosion and was demolished by the Air Force in December 1976.

(Right) Launch Complex 5 is now part of the Air Force Missile and Space Museum. This Mercury Redstone replica stands on the exact spot where Alan Shepard and Gus Grissom were launched on 15 minute suborbital trajectories back in 1961.

(Left) The launch mount and flame deflector of pad 19 for Gemini-Titan is seemingly well preserved after nearly forty years of exposure to the elements after the last Gemini launch in November 1966.

(Right) The distinctive runoff basin at pad 19, characteristic of the Martin Titan pads at Cape Canaveral.

(Left) The skeleton of the unique erector gantry still resides atop the ruins of pad 19 (the launch mount is at right). The white room structure at left (where the Gemini astronauts entered their spacecraft) has been removed from the pad and restored by volunteers. It now occupies a place of honor at the Missile and Space Museum.

(Right) Yesterday and Today: The distinctive concrete pillars of pad 34, location of the tragic spacecraft fire that took the lives of the Apollo 1 crew in January 1967, frames the present-day Delta IV launch facility at Complex 37. Note the 'Abandon in place' stencil on the nearest pillar. Pad 34 was also the site of the first manned Apollo launch in October 1968.

(Left) Sunlight streams in through the launch mount ring of pad 34, the only equipment that remains intact at the site. The umbilical tower and mobile service structure were dismantled in the early 1970's.

(Right) The Mercury 7 Memorial at the entrance to Complex 14 at Cape Canaveral. The monument was fashioned by Convair Astronautics from the same thin-gage stainless steel that was used to construct Atlas rockets. A buried time capsule at the site (emplaced in 1963) is intended to be opened 500 years after burial.

(Left) The magnificently restored Gemini White Room from pad 19 is now on permanent display in the Missile and Space Museum at the Cape on the site of former Launch Complex 26 on the south end of the base.

Delta 190 soars aloft from pad 17A as seen from the Space and Missile Museum on December 11, 1989. The missiles in the foreground are Snark (left), Bomarc, Polaris and Mace.

The gate guardian at the South (main) entrance of Cape Canaveral is a magnificently restored Navaho missile.

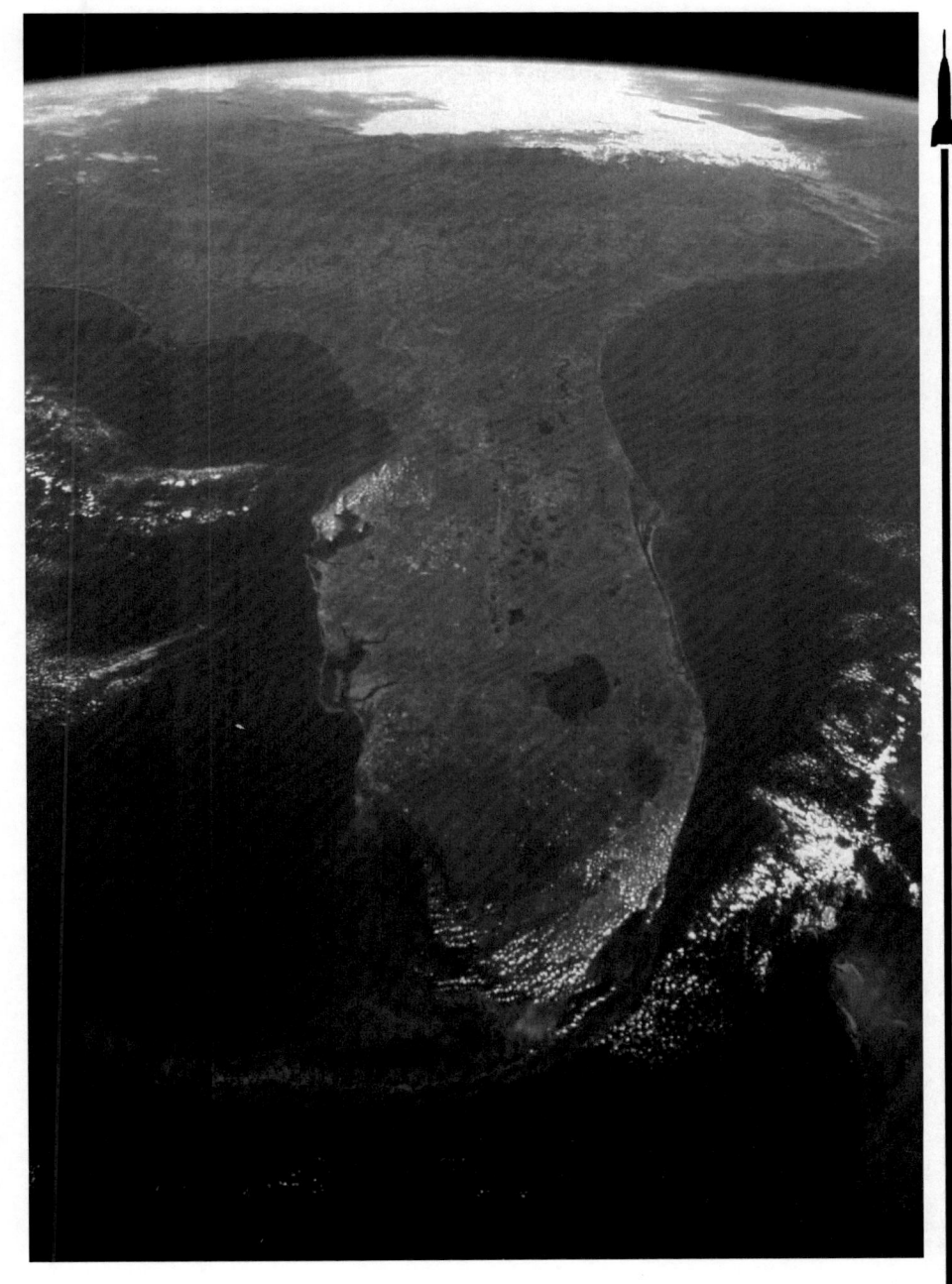

Cape Canaveral is the prominent point of land on the west coast of the Florida peninsula (right) in this orbital portrait taken from the STS-95 shuttle mission in October 1995.

The famous 'ICBM Row" at Cape Canaveral looking north from Complex 36 in this photo from August 1963. the Industrial Area is located in the far upper left hand corner.

Florida experiences some of the most intensive lightning activity in the world, as evidenced in this night time portrait of a Jupiter missile at Launch Complex 5/6 in 1959.

The Building Directory sign at the entrance to the Industrial Area at Cape Canaveral in 1958. The missile on top is a Lark.

The relative simplicity of the Thor launch facilities at pad 17B is evident in this picture of Thor 121 shortly before launch on April 19, 1958 (test 858).

The final Navaho missile lifts off from the pad 9 hardstand on November 18, 1959 (test 1811). Note the unusual Navaho erector gantry in the 45 degree parked position for launch, at right.

Rising on a pillar of fire, Thor 114 ascends straight and true from pad 17A on January 28, 1958. The flight was rated a partial success.

Classic portrait of an early Atlas A test vehicle on the pad at Cape Canaveral in early 1958. Atlas A vehicles were adorned with these tracking stripes beginning with the third test flight.

Propellant loading has begun at pad 26A on the historic night of January 31, 1958 as Juno I vehicle UE (missile 29) is prepared to launch the first American earth satellite, Explorer 1.

Vanguard TV-4 is prepared for the Vanguard 1 launch in March 1958 at Launch Complex 18A. In the background is the gantry and tactical launch pad for Thor (stub tower at left), Complex 18B.

Aerial portrait of the Navaho area at Cape Canaveral in late 1956, with the Industrial Area (to the west) visible at the horizon. Navaho's pad 9 lies under the shelter at left center, while the concrete apron for the Navaho tactical site (pad 10) is located at right. A Navaho vehicle is just barely visible in front of the weather shelter.

Convair's Complex 13 for Atlas is pictured in 1958 from just off shore, looking across the Cape to the Banana River. The hangars of the Industrial Area are visible in the upper right corner.

Titan II vehicle N-32 is stationed within the erector gantry at pad 15 prior to launch on February 26, 1964. The 102 foot umbilical tower is positioned at the right-hand side of the vehicle.

The welcome home ceremony for astronaut John Glenn at Cape Canaveral on February 23, 1962 at Hangar S. Greeting America's newest hero was President Kennedy, Vice-President Johnson and NASA administrator James Webb. Glenn's Friendship 7 capsule is visible at right.

Gordon Cooper rides the final Mercury-Atlas vehicle to orbit (MA-9) on May 15, 1963. Nothing captured the attention of the general public more than the manned space flights from Cape Canaveral in the 1960's.

The third Atlas-Centaur vehicle is photographed from the service tower at pad 36A in June 1964. Note the angled flame deflector at the base of the pad, and the Cape lighthouse is visible in the background.

The second-to-last Titan II missile, N-33, rises from pad 15 at the Cape on March 23, 1964. The first Gemini-Titan vehicle can be glimpsed inside the pad 19 erector at left. The 310 foot gantry at Complex 34 often served as an observation point for missile launches at Cape Canaveral.

The first Air Force Titan IIIC launch vehicle is in position for flight at pad 40 of the new Integrate, Transfer and Launch (ITL) facility at Cape Kennedy in late May 1965.

The twelfth and final Gemini-Titan rocket ascends past the surf near pad 19 at Cape Kennedy at 3:46 pm on November 11, 1966 with astronauts Jim Lovell and Buzz Aldrin aboard.

The first manned Saturn IB rocket lifts off from pad 34 at Cape Kennedy on October 11, 1968 at 11:04 am EDT. Apollo 7 was also the first American space mission after the devastating Apollo spacecraft fire in January 1967.

The fifth test launch of the Navy's Trident II (D-5) missile lifts off from Complex 46 on July 20, 1987. Trident II was the last missile to be flight tested at the Cape.

The debut of the Titan IV heavy-lift booster on June 14, 1989 from pad 41 at Cape Canaveral continued the rebirth of the American expendable launch vehicle industry, after the production lines had been closed prior to the Challenger shuttle disaster on January 28, 1986.

An Atlas IIAS rocket (AC-135) with a commercial payload roars off pad 36B in October 1997. Atlas IIAS (S for Solids) was the first Atlas variant to utilize solid rocket motors.

The Cassini space probe's fiery ascent on a Titan IVB rocket from pad 40 was one of the most beautiful launches at Cape Canaveral in recent memory. Cassini's long journey to Saturn began on October 15, 1997 at 4:50 am EST, with arrival at the Ringed Planet nearly seven years later on July 1, 2004.

NASA's first Mars rover Spirit (MER-A) is concealed beneath the payload fairing of Delta 298 at pad 17A. The service tower is being rolled back prior to liftoff on June 9, 2003.

Three of four Titan sites are visible in this off-shore aerial portrait of the Gemini 1 launch on April 8, 1964 from pad 19. Pad 16 is to the left of pad 19, and pad 20 is to the right.

A shrouded Atlas A missile (right) and its escort convoy pause on Merritt Island on the way to the causeway leading to the main gate of Cape Canaveral. This photograph was taken some time in 1957.

An aerial portrait of Lockheed Martin's "clean pad" for Atlas V, Launch Complex 41 at Cape Canaveral, with the launch vehicle for the Pluto New Horizons probe in January 2006. The rocket is usually moved to the launch pad from the Vertical Integration Facility 24 hours or less prior to the scheduled launch time.

Boeing's Delta IV launch site, LC-37, is a more traditional setup than for Atlas V, with vehicle processing occurring mostly at the launch pad.

The Navy's Polaris pads were located at the south end of the base (pad 29 and 25 are at the lower left), with the Army's Pershing pads (Complex 30) located just above them. Complex 26 (soon to be the missile museum) is visible at lower right in this photograph taken by the Air Force in 1964.

The flat Cape Canaveral terrain is emphasized in this clearing at the Army's Vertical Launch Facility in August 1959. At left is the ill-fated Juno II vehicle AM-16 on pad 5, with Jupiter missile AM-15 on pad 26B at right, cocooned within the new Noble Co. service structure.

CAPE CANAVERAL AIR FORCE STATION
DRIVE-THRU TOUR
Sunday 9:00 A.M. - 3:00 P.M.

FOREWORD

The Eastern Test Range is part of the Eastern Space and Missile Center under the Air Force Systems Command. It is the final checkpoint for programs planned and developed by military and civilian engineers and scientists for space exploration and national defense.

The Eastern Test Range is a vast test complex that includes administrative headquarters at Patrick Air Force Base, Cape Canaveral Air Force Station 15 miles north of Patrick Air Force Base, and the remainder of the Eastern Test Range which extends more than 10,000 miles into the Indian Ocean. This includes island stations at Grand Bahama Island, Grand Turk, Antigua, and Ascension Island.

Not only the Air Force, but the Army, Navy, NASA, and friendly foreign powers rely upon the support services of the Range for their missile and space projects.

Missile tests are conducted under the most exacting laboratory conditions. Each missile's performance is scrutinized by instruments so varied, so sensitive, and so positioned that throughout the flight it is under continuous surveillance. Radar and optical devices track each rocket as it streaks across the sky while telemetry equipment detects and records vital information being transmitted by tiny sensors planted throughout the missile's systems.

Through the skillful use of this ultra-sophisticated equipment, missile and space systems can be perfected with a minimum number of flights.

This booklet describes what you will see at Cape Canaveral Air Force Station. A map showing the Cape Canaveral Air Force Station tour route is presented in the center of this booklet. Cameras are permitted along the tour route including the stop at the Space Museum.

Welcome to the Cape

Welcome to Cape Canaveral Air Force Station. It is our pleasure to be your host on this historic site where America's achievements in missile and space exploration had their beginnings.

As you visit ICBM row, the launch pads, and the Air Force Space Museum, you will be seeing the evidence of our national commitment to the development of intercontinental missiles and the conquest of space. You will appreciate the enormity of the technical challenges that were met by the pioneers who freed us from the planet Earth.

We sincerely hope that you will share our pride in the dedicated men and women, past and present, whose accomplishments have kept America in the forefront of space science. Their work is bringing untold benefits to all mankind.

We are pleased that you are visiting with us and hope you have a most pleasant and educational stay.

NOTE: The Drive-Thru Tour of the Cape Canaveral Air Force Station is about 20 miles in length. May we suggest that you check your gas gauge before embarking on this space adventure?

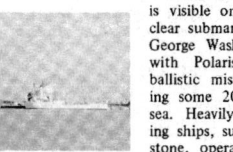

From Gate 1, Port Canaveral is visible on the right. Here nuclear submarines such as the USS George Washington are equipped with Polaris or Poseidon fleet ballistic missiles for test launching some 20 to 30 miles out to sea. Heavily instrumented tracking ships, such as the USNS Redstone, operate out of this deepwater port.

CHECKPOINT A

Inside the gate, just before you turn left, you will note a new facility for Trident submarines and its tall port service crane. After you have turned left and start north, you will see a large antenna field ahead and to your right. Each antenna serves a specialized purpose in the communications system which ties the Cape to the downrange stations, instrumented ships and aircraft. The Eastern Test Range has an around-the-world communications capability.

Continuing north, you will see, on the median in the center of the road, a white, winged missile with a sign pointing to the "Space Museum." This missile is a Hound Dog. It is launched from the wing of a B-52 bomber and has a range of approximately 700 miles at supersonic speeds. This Hound Dog is pointing to the Museum, indicating you should make a RIGHT turn on Lighthouse Road after passing the missile.

CHECKPOINT B

After you turn toward the Museum, you will notice some square buildings to the right through the foliage. In this "spin test" facility, satellites are given a final evaluation before being moved to the Delta pads. Delta space vehicles are launched from the two gray structures you see straight ahead as you proceed toward the Museum. From these two towers on Complex 17, Delta has sent various satellites into space; this is the most active launch area at either Cape Canaveral Air Force Station or the Kennedy Space Center. From here most of NASA's astronomy, biosatellites, communications, navigational and weather satellites have been prepared and launched.

These include Echo, Syncom, Relay, Telstar, Early Bird, Intelsat, Explorer, Tiros, Pioneer, Skynet, Symphonie, Telesat, HEOS, Westar, SMS, Marisat, Palapa, Satcom, NATO, GEOS, GMS, SIRIO, OTS, ISEE, Meteosat, CS, and IUE.

CHECKPOINT C

On your right and left are signs pointing to the entrance of the Air Force Space Museum. Turn RIGHT at the signs and follow the marked road through the Museum grounds to the parking area. You are now on Complexes 26 and 5/6, both inactive. After you have parked your car, feel free to walk around the grounds and visit the buildings at Complex 26. The 40-acre area is the birthplace of America's manned and unmanned space exploration.

From Pad A on Complex 26, Explorer I, the first U.S. satellite to be successfully placed in orbit, was launched on January 31, 1958. Astronauts Alan B. Shepard and Virgil I. "Gus" Grissom in 1961 were launched into suborbital flight in Mercury spacecraft by Redstone rockets from Complex 5/6. The Air Force designated these sites as Museum grounds in 1963 to preserve the tradition and history of rocketry for the American public. Dozens of actual missiles and rockets are displayed on the grounds. These range from a German V-1 "buzz bomb" to a Mercury-Redstone and Atlas and Titan intercontinental ballistic missiles. The displays are multi-Service and include manned, military, and scientific boosters.

Most of the missiles and rockets displayed on the Museum grounds were developed through launches from Cape Canaveral and flights down the Eastern Test Range. One of the missiles on display, the Snark, was actually flown down the Range and returned to land on the Cape's "Skid Strip," now used as an aircraft runway. In the exhibit hall, displays include actual missile warheads, nose cones, fuel and oxidizer pumps, thrusters, weather payloads, engine cutaways, space foods, and numerous satellite and rocket models. Visitors are welcome to take photos of displays. Airmen on duty will be happy to answer your questions.

Leave the Museum by the same gate you entered. When you reach the Museum entrance sign, turn LEFT, drive approximately 100 yards and turn RIGHT (north). You will pass a pump, or deluge, station on the left. This facility is used to pump thousands of gallons of cooling water over the pads of Complex 17 during liftoff. On the right you will pass the inactive Air Force Blue Scout and the Navy Vanguard pads. Travel straight ahead to the intersection, then turn RIGHT.

CHECKPOINT D

You are now driving toward the USAF's Minuteman Complex. The three-stage Minuteman is an intercontinental ballistic missile which is propelled principally by solid fuel and has a range of 6,300 miles. This missile was launch-tested primarily from 90-foot deep "silos" not visible from the road. The only visible evidence of the Minuteman's testing is one short red service tower and two sandbagged blockhouses that resemble beehives. Turn RIGHT as you approach the entrance to the Minuteman Complex and follow the road as it turns left along the fence. Minuteman is now operational with SAC, the Strategic Air Command, and is America's primary intercontinental ballistic missile.

CAPE CANAVERAL AIR FORCE STATION

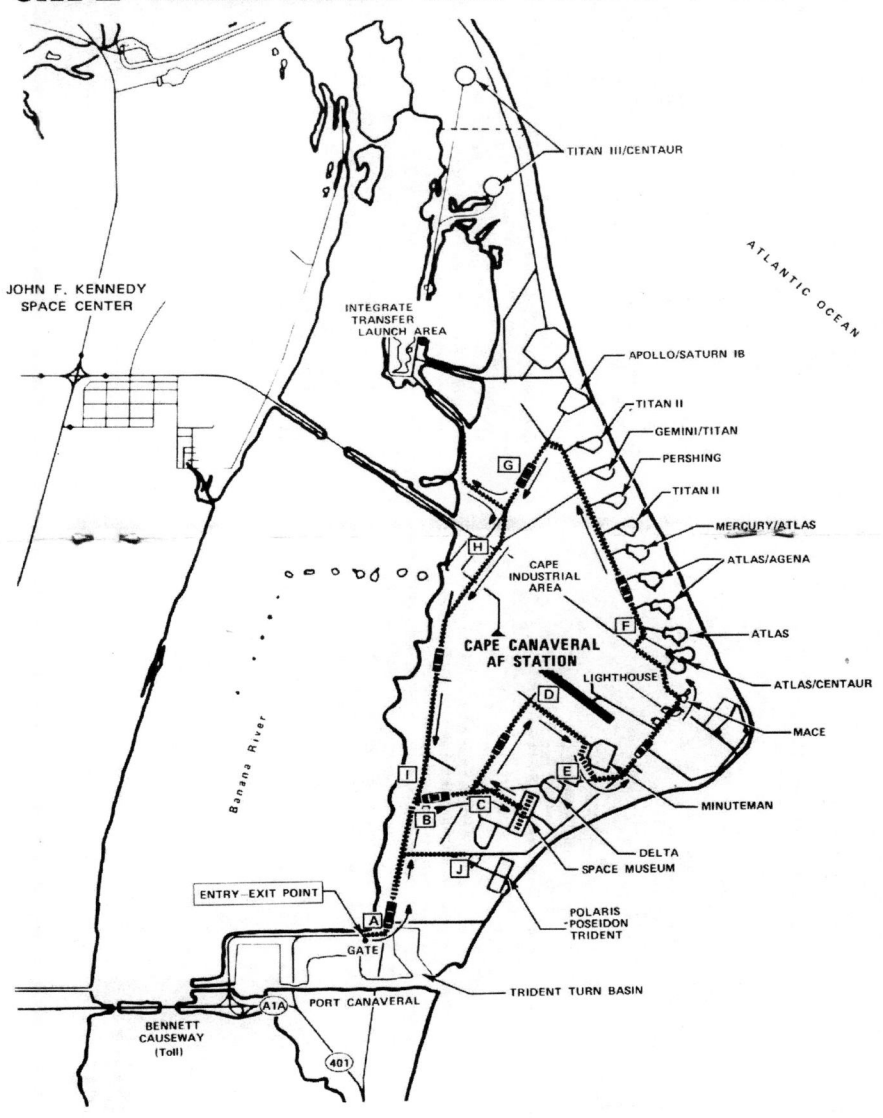

CHECKPOINT E

Off to your right you will see the Delta pads which you saw from the Museum. When you reach the intersection, turn LEFT.

Drive about a quarter of a mile to a stop sign at the intersection of Lighthouse and Pier Roads. Turn LEFT and proceed north. You are approaching the scene of our earliest missile launches, some of them dating back to the early 1950's.

On the left, and very prominent, is a 165-foot black and white lighthouse. A lighthouse has been a Cape landmark since 1868. The first lighthouse was built of wood but was later rebuilt closer to the ocean with riveted steel plates lined with brick. Still later it was disassembled and moved one mile to this site. From this distance the structure greatly resembles a missile and some hapless newcomers to the Cape have mistakenly taken it for one.

On the right, the building with the sloping green roof is the simulated hardsite used to launch the Mace missile. The two large pipes coming from the lower rear portion of the hardsite are blast deflectors. The Mace as well as other aerodynamic cruise missiles were built to fly in the atmosphere and required air to support combustion in their engine, much like pilotless aircraft. In contrast, ballistic missiles, carrying their own oxidizers, can fly beyond the atmosphere into space.

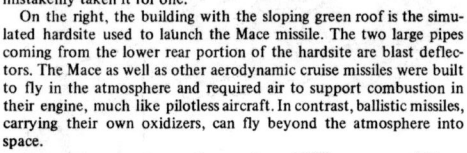

As you bear LEFT, on your right are two service towers constructed by NASA for its Centaur launch vehicle at Complexes 36A and 36B. Centaur, launched by an Atlas first stage, was the first United States rocket to employ liquid hydrogen fuel. Missions launched by Atlas-Centaurs include Pioneers 10 and 11 and Mariner 10. Pioneer 10 was launched March 2, 1972, on a trip to the vicinity of the planet Jupiter, arriving there December 3, 1973, and then continuing on out of the solar system. Pioneer 11 swept by Jupiter December 3, 1974, reached the orbit of Saturn in 1976 and should reach the orbit of Uranus in 1979. Mariner 10 was launched November 3, 1973; in early 1974 it successfully swung by the inner planets Venus and Mercury. In late 1974 and early 1975 Mariner 10 flew again past Mercury.

CHECKPOINT F

Following the tour signs you will complete a sharp RIGHT turn. Ahead and to your right are sites of four Atlas complexes. The first was Complex 11, one of several launch areas operated by the Air Force's 6555th Aerospace Test Group. Atlas, our first ICBM, has become obsolete as a weapons system. It was converted to a manned and satellite booster, lofting Mercury astronauts and satellites into orbit and many payloads into deep space.

Again, on the right, Complexes 12 and 13 supported Atlas-Agena programs which were concerned primarily with deep-space exploration. Five Mariner spacecraft were launched from Complex 12 and 13 to fly past the planets Venus and Mars. The Moon-exploring Rangers were also launched by Atlas-Agenas.

Ahead, to the right, is Complex 14 from where all of our Earth orbiting Mercury Astronauts were launched. Its structure, which was familiar to millions who watched Mercury-Atlas liftoffs on television, was dismantled in 1976. Also on the right, you will pass a monument to the original seven astronauts. The design incorporates a 13-foot high astronomical symbol for the planet Mercury and the number 7 for the original seven U.S. astronauts. There is a time capsule buried beneath the monument which is to be opened in the year 2464. It contains reports, photographs, motion pictures and other memorabilia.

Farther along, on the right you see Complex 15, which is the beginning of a series of Titan II complexes. Titan II, like the Atlas, is a liquid-fueled heavyweight. The Titan II test program has been completed for some time, and like Minuteman, Titan II is in the Strategic Air Command arsenal. Complex 16, converted to a service area for Apollo ground support equipment in January 1965, was returned to the Air Force in June 1972. It is now used as an operational launch site for Army Pershing missiles which are launched under simulated combat conditions.

The 103-foot Titan II was also adapted to the role of space booster.

The next complex on the right is Complex 19. While similar to Complexes 15 and 16, it was modified for the Gemini Program which sent two-man crews into Earth-orbital flight. Titan II served as the booster in the program. There were ten manned Gemini missions in 1965 and 1966.

Complexes 34 and 37, built for NASA's Saturn I and IB launch vehicles, were located ahead to the right but have been dismantled. The Saturn IB developed 1.6 million pounds of thrust and was capable of placing an Apollo spacecraft, carrying three astronauts, in Earth orbit. Blockhouses for the two complexes remain.

Employing Saturn I and IB vehicles, NASA flew a number of missions from these pads with the Apollo spacecraft. All were unmanned, except the final launch in the series, Apollo 7, which was the first U.S. three-manned mission. Three Saturn IB launches in the Skylab Program and one for the Apollo Soyuz Test Project were conducted at Launch Complex 39B, KSC, NASA.

CHECKPOINT G

To view the Titan III launch complex, turn RIGHT on Titan III Road. After driving one mile, you will reach Area 60A. STOP at this point.

As you survey the Titan III area, the first large building to enter your view is the Vertical Integration Building (VIB). The modified Titan II core, or booster, is taken to this location on arrival at the Cape, erected to an upright position, and mated with its upper stage, either a transtage or a Centaur. Farther north, the second large building is the Solid Motor Assembly Building (SMAB). After a complete checkout in the VIB, the Titan booster is moved on rails to the SMAB, where two solid motors are attached, one on each side of the core vehicle. These give the vehicle 2.4 million pounds of thrust at lift-off. After the solid motors are attached, the integrated booster is moved to one of the two launch structures which can be seen in the distance. Titan IIIE/Centaurs were launch vehicles for the twin 1975 Viking missions to Mars. Two Voyager missions to conduct exploratory investigations of Jupiter and Saturn were successfully launched by Titan IIIE/Centaurs in 1977. Turn around at Area 60A and return to Cape Road, then turn RIGHT and proceed toward the Industrial area.

At the intersection of Cape and Hangar Roads, bear LEFT on Cape Road and continue past the USAF Range Control Center, the nerve center for the entire Range. It is from here that all Cape launch and tracking activities are directed.

Note: At the intersection just beyond the Range Control Center, continue straight ahead to Gate 1. (Due to Space Shuttle operations, no drive-thru tour is available at the Kennedy Space Center, NASA.)

CHECKPOINT H

On the right and left as you travel toward Gate 1, you will see hangars and support buildings used by the Air Force, NASA and their contractors. These facilities are used mainly as supply warehouses, laboratories, administrative offices and shops. Evidence of the self-supporting nature of this area is the fire station you will note on your left. To your right is the office of the Commander, Cape Canaveral AFS, who could be compared to your local mayor.

When you are again on the four-lane highway, you will soon pass an area where various solid, liquid, and gaseous fuels are stored. In the solid propellant storage area you will see a square building atop a mound, marked NDTL, which houses one of the world's largest "X-rays," the betatron, used to seek out flaws in solid propellants. All fuels which are stored in this area must be treated as high explosives.

CHECKPOINT I

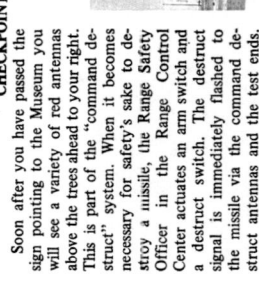

Soon after you have passed the sign pointing to the Museum you will see a variety of red antennas above the trees ahead to your right. This is part of the "command destruct" system. When it becomes necessary for safety's sake to destroy a missile, the **Range Safety Officer** in the **Range Control Center** actuates an arm switch and a destruct switch. The destruct signal is immediately flashed to the missile via the command destruct antennas and the test ends.

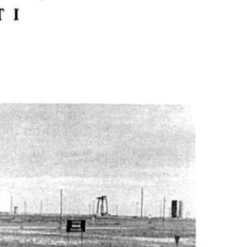

Naturally, the missile is allowed to fly as long as safety considerations permit, because it continues to transmit valuable information — perhaps even explaining why it malfunctioned.

CHECKPOINT J

To see the Navy's Polaris/Poseidon/Trident launch pads, turn LEFT on Pier Road. Navy missiles are assembled here rather than in the hangars at the Industrial Area because of the convenience to Port Canaveral. Turn around at the spot indicated and return to Cape Road, then turn LEFT to reach Gate 1.

The Air Force hopes you have enjoyed your tour of Cape Canaveral AFS, and welcomes comments and suggestions. Please address comments to ESMC(PA), Patrick Air Force Base, Florida, 32925.

After you leave through Gate 1, you will approach State Road 528. A RIGHT turn will take you to Merritt Island and U.S. 1 on the mainland. Ocean beach communities can be reached by turning LEFT on A1A.

Travel safely, and please come back!

From the vantage point of a berm located at the south end of the base, a Delta II rocket stands ready at distant pad 17A (the third tower from the left on the horizon). The occasion is the launch of the first Mars Rover "Spirit" in June 2003. The bulge of the Cape can just be glimpsed to the right of the missile assembly building at right.

CAPE LORE

An unusual launch abort: Most rockets in the early days of Cape Canaveral reacted severely to the shutdown of an engine or a bout of combustion instability at the moment of liftoff: they blew up, destroying the launch stand in the process (see "Incidents and Accidents"). Not so the Titan II. When vehicle N-4 experienced combustion instability during ignition of the first stage engines on June 27, 1962, the launch sequencer commanded engine shutdown at T - 2.5 seconds. This saved the vehicle and the launch stand. There was a "severe transient" in thrust chamber number 2 that triggered the automatic cutoff. The shock wave was so powerful that it sheared off the engine nozzle and hurled it several hundred feet from the launch pad, where astonished Martin engineers discovered it after the vehicle was saved. The engine set was replaced right at the pad and N-4 was launched on July 25,1962. There was no more combustion instability, but the second stage malfunctioned and the re-entry vehicle failed to separate. (from "Titan II," by David Stumpf, University of Arkansas, 2000)

Scrub and de-scrub: Cancellations and launch scrubs are part of the rocket business. It is no less true in 2006 than it was in 1966, when a scrub was actually rescinded and the launch was allowed to proceed for the first time in Cape Canaveral history. The occasion was the maiden launch of the Saturn IB rocket on February 26, 1966. NASA designated the launch AS-201, and to the Air Force it was known as Test 0195 on the range. After a troubled checkout on newly refurbished pad 34, AS-201 appeared ready to fly as the countdown entered the final minute. At T - 35 seconds a high pressure purge of first stage components with nitrogen gas was commanded by the sequencer, but pressure in the compressed nitrogen spheres dropped rapidly below the redline figure and the count was halted at T - 4 seconds. Ground controllers investigated various "work-arounds" and concluded that any corrective action would take longer than the launch window allowed, and Test 0195 was scrubbed. A few of the ground controllers disagreed, and they managed to convince NASA's launch director to allow them to run a simulation to demonstrate that the nitrogen pressure would remain adequate until first stage burnout. The simulation and backup calculations demonstrated that the nitrogen pressure would hold, and mission managers agreed to resume the countdown. The count was recycled to T - 15 minutes and AS-201 lifted off smoothly at 11:12 am on February 26. Launch vibrations tripped all of the high voltage circuit breakers at the pad and blacked out the Cape's power supply two seconds after liftoff. Launch controllers were literally in the dark (and without trajectory data for Range Safety), but the first "un-scrub" in Cape history was successfully pulled off by the Saturn launch team. The Apollo spacecraft survived the suborbital flight and splashed down in the Atlantic Ocean 37 minutes after liftoff where it was recovered by the carrier USS Boxer. (from "Moonport," by Benson and Faherty, NASA SP-4204, 1978)

The Inter-Pad Missile The Redstone rocket was known as 'Old Reliable' in the missile business fifty years ago, but on the night of October 30, 1956 a malfunction aboard missile 25 (code US) almost resulted in the destruction of a competitor's launch site and earned the Army an awful lot of ribbing from their opposite numbers at the Cape. Missile number 25 was a Jupiter A, the slightly disingenuous name for a Redstone vehicle that carried experimental components for the Jupiter IRBM. A poorly soldered connection in the yaw gyro was severed by the heavy vibration of the missile at liftoff from pad 5, and the Jupiter A immediately flew out of control at 10 seconds and leveled off no more than 300 feet off the ground. The missile headed straight for the brand-new Thor launch complex right next door to pad 5. The Range Safety Officer immediately destroyed the rocket, which crashed to the ground in an empty field between the launch sites. The teasing of the Army launch team was merciless. The Army was accused of having a very inaccurate missile that was incapable of hitting the target even at extremely short range (Thor was still two months away from its first launch attempt at the time). In the official records, the radial dispersion of missile 25 (an indication of how far the warhead missed the target) was listed as 264,900 meters! For the near miss on the Thor pads, the abbreviated flight of missile number 25 became known at Cape Canaveral as the Inter-Pad Missile. (from "Countdown to Decision," by John B. Medaris, Putnam,1960)

Night Watch at the Atlas-Agena pad: Billy Wohlforth was a welder who worked for Lockheed Company at Cape Canaveral for over twenty years. He started with Lockheed in the Industrial Area in October 1961 engaged in the Ranger lunar probe program at Hangar E. He recalled a thrilling incident that he experienced while standing guard at the Atlas-Agena launch pad one night in 1964.

During night watch the guard was all alone in the gantry, and to keep awake Wohlforth would walk from one level of the gantry to another. "All of a sudden the structure began a tremendous shaking. It really scared me because I thought the missile had ignited. A second later I realized that it was a Minuteman missile launch (from the nearby Minuteman test silos). I rushed out onto the catwalk and looking up I could see directly into the tail of the missile as ashes fell all around," he said. The night watch would never be the same after that experience. (from the LOCKHEED STAR GAZER, September 1992) A similar incident nearly twenty years later would give a Space Shuttle technician the scare of his life as he sat in the cockpit of the shuttle Columbia during a night time test session. Without warning the shuttle began to shake violently as a bright orange light filled the cockpit windows. Fortunately the shuttle was not launching of its own volition: an unannounced launch of a Titan IIIC rocket from just across the boundary line at Cape Canaveral caused the shuttle to behave as if it was about to lift off.

The Jungle Missile: The Snark was a large subsonic aerodynamic missile that had a dubious record of reliability at Cape Canaveral in the 1950's. "Snark-infested waters" was the oft-repeated joke about all the Snark missiles that crashed off shore from the Cape before making it back to the Skid Strip for recovery. The missiles were supposed to make a turn back towards Florida after traveling downrange, but on December 5, 1956 the Snark missile bearing the Air Force serial number 53-8172 not only failed to make the turn for home, but it headed straight for the east coast of Brazil before disappearing from the Cape's radar screens. That was the last anyone heard from the runaway missile until a report reached the Pentagon 26 years later that a farmer had discovered the wreckage in the Brazilian jungle in 1982. It turns out that the story is unconfirmed (a writer found conflicting versions of the story that was first reported by Reuters news agency in January 1983. A lost plane (or missile!) always makes a great story...

The Perfect Souvenir: The U.S. Air Force took back control of Launch Complex 13 from NASA at Cape Canaveral in 1968 in order to conduct a series of classified space missions with Atlas-Agena boosters. All but one of these missions made it into orbit: on December 4, 1971 an Atlas-Agena rose off the launch pad in the late afternoon, but instead of reaching orbit the vehicle malfunctioned and was destroyed by the Range Safety Officer. Pieces of debris were recovered for the accident investigation, which later became "Perfect Souvenirs" when an unknown entrepreneur fashioned pieces of the debris into tie tacks. The keep-sakes were made available to the public (or as a hand-out for contractors at the Cape) in 1972.

"Launch This Way!" It was official NASA policy that every component of the Project Mercury spacecraft (no matter how small) should be stamped with the project's Mercury astronomical symbol. General Dynamics Astronautics, manufacturer of the Mercury-Atlas launch vehicle, instituted their own tradition of affixing the General Dynamics 'R' for Reliability sticker on every Atlas built for the Mercury program. When the vehicle for the final Mercury mission was rolled out of the factory in a ceremony at San Diego on March 15, 1963 (Atlas 130D), MA-9 astronaut Gordon Cooper and his backup Alan Shepard signed the General Dynamics sticker for good luck before it was applied to the Atlas (photo). Cooper also wrote "Launch This Way!" with an arrow pointed forward on the skin of the Atlas. NASA evidently followed his advice on May 15, 1963 with a nearly flawless launch for Cooper and his Faith 7 spacecraft.

Titan Mission Patches: Martin Marietta and Lockheed Martin employees (and their military counterparts at Cape Canaveral) developed an interesting tradition in the late 1980's to give an identity to the Titan 34D or Titan IV launch vehicle that they were currently working on. Every vehicle was given a colorful nickname (at least until the more serious brass at the Pentagon put an end to the frivolous nicknames) and the art department designed an unofficial mission patch for each launch. Available at the Lockheed Martin company store at Cape Canaveral, the patches (and also pins) were worn by military personnel on their uniforms and by the contractors on their jackets or other clothing. The nicknames ranged from cartoon characters (Bart Simpson, Spaceman Spiff among others) to those honoring significant Air Force personalities (such as Gus, for astronaut Gus Grissom). The patches were also produced for Titan IV missions at Vandenberg AFB. The patch tradition was maintained until the final Titan launch at the Cape in April 2005, but does not seem to have been adopted for the new Atlas V rocket.

The Day Gordo Cooper Buzzed The Cape: Astronaut Gordon Cooper flew the sixth and final manned flight of Project Mercury on May 15-16, 1963, but an incident at the Cape a few days before the launch almost cost him his ride into space. Angry at a last-minute, unauthorized change to his space suit for the flight, Cooper took off from nearby Patrick Air Force Base in the F-102 jet fighter that the astronauts had at their disposal. After putting the jet through a series of aerobatic maneuvers, Cooper dove toward the Industrial Area of the Cape and decided to buzz the Project Mercury headquarters at Hangar S at very low altitude (some say he also buzzed his Atlas booster at pad 14). Mercury Operations Manager Walt Williams, the number three man at NASA, happened to step out of a doorway just as Cooper roared by, and was so startled that he dropped a stack of papers he was carrying (but according to Chris Kraft, Williams told him he was in his office in Hangar S when Cooper raced past the building at window level). Livid with rage at the audacity of Cooper's stunt, Williams demanded that Cooper be removed from the MA-9 flight and be replaced by his backup, Al Shepard. Williams reportedly did not relent until around 10 o'clock that evening, allowing Cooper to keep his seat on the flight after an agonized evening of uncertainty. Cooper was later told by Deke Slayton that President Kennedy personally intervened with the NASA hierarchy on his behalf. Cooper noted that it was good to have friends in high places. (from "Leap Of Faith," Gordon Cooper with Bruce Henderson, Harper Collins, 2000)

The MA-9 sticker signed by Cooper and Shepard is carefully applied to Atlas 130D by a General Dynamics worker on March 15, 1963.

The "Oscar" patch depicted above was produced for Titan IV mission K-20 (Titan IVA number 17) that was launched on November 8, 1997 from pad 41.

These anecdotes are representative of the many stories of life and work at Cape Canaveral that can be found in the large number of space books published in the past ten years, and available in the old classics from the 1950's and 60's. Check for these titles at on-line booksellers like Amazon Books or visit your local library.

BIBLIOGRAPHY: Books About Cape Canaveral

Martin Caidin, "Spaceport U.S.A.," E.P. Dutton, 1959

William Roy Shelton, "Countdown, The Story of Cape Canaveral," Little, Brown, 1960

Mel Hunter, "The Missilemen," Doubleday.. 1960

Charles Coombs, "Gateway to Space," William Morrow, 1960 (juvenile)

C.W. Scarboro and Stephen Milner, "Cape Kennedy, America's Spaceport," Pioneer Press, 1966

L.B. Taylor, Jr., "Liftoff! The Story of America's Spaceport," E.P. Dutton, 1968

ESMC Pamphlet 190-1, "From Sand to Moondust," Eastern Space and Missile Center, 1986

Mark C. Cleary, "The 6555th, Missile and Space Launches Through 1970," 45th Space Wing History Office, Patrick Air Force Base, FL, 1991

Mark C. Cleary, "The Cape, Military Space Operations 1971-1992," 45th Space Wing History Office, Patrick Air Force Base,. 1994

Cliff Lethbridge, "Cape Canaveral, 500 Years of History, 50 Years of Rocketry," Space Coast Cover Service, 2000

Guenter Wendt," The Unbroken Chain" Apogee Books 2001

A crowd of spectators gather just south of pad 34 at Cape Canaveral to watch the launch of Apollo 7 on October 11, 1968. This was the first manned Apollo test flight.

PHOTO CREDITS

Boeing
p.45 (lower right); p.263 (2)
Courtesy of British Interplanetary Society
p.6
General Electric
p.94 (upper)
Peter Gualtieri
p. 298, 299
Courtesy of Al Hartmann
p.29 (USAF); p.104 (USAF, lower) p.122 (North American Aviation), p.123 (North American Aviation, top left, right); p.282 (USAF)
Bart Hendrickx
p.138 (2)
Courtesy of Ed Hengeveld
p.141 (NASA, center, lower left and right); p.219 (NASA, lower left); p.221 (NASA, upper); p.222 (NASA, left and lower right); p.228 (NASA, 2); p.294 (NASA); p.295 (NASA); p.317.
Courtesy of Peter Hunter
p.123 (lower); p.125 (inset); p.281; p.283; p.287 (lower)
Jet Propulsion Laboratory
p.60 (lower); p.96(left); p.97 (2); p.155 (upper left, right); p.156 (lower); p.279; p.289
Lockheed Martin
p.69 (lower right); p.242; p.246 (lower right); p.247 (2); p.255 (lower left); p.316 (lower)
National Aeronautics and Space Administration (NASA)
Front Cover; p.3; p.11 (lower); p.21; p.23 (2); p.32 (lower); p.35 (3); p.36 (upper); p.40 (upper); p.51; p.64 (left, lower right); p.73 (left); p.108 (center; lower); p.110 (upper); p.113 (lower); p.114 (3); p.115 (lower); p.118 (left); p.119 (4); p.120 (upper); p.124 (upper right and center); p.127 (center and lower right); p.130 (center; lower); p.131 (3); p.132 (lower right); p.133 (3); p.134 (3); p.135 (3); p.137 (lower left; right); p.142 (2); p.143 (lower); p.144 (lower right); p.145 (upper); p.147 (upper right, lower left); p.150 (upper); p.151 (right); p.152 (3); p.153 (2); p.154 (upper); p.158 (lower left; right); p.160 (left); p.161 (lower left); p.165 (lower left; right); p.166 (3); p.168 (left); p.178 (lower); p.179 (3); p.180 (3); p.181 (3); p.182 (3); p.183 (2); p.184 (upper; lower right); p.185 (3); p.186 (left; lower right); p.187 (upper; lower left); p.188 (2); p.189 (upper right); p.193 (upper left; right); p.195 (upper); p.196 (2); p.199 (lower); p.200 (upper; lower); p.201 (3); p.202 (3); p.204 (3); p.205 (3); p.212 (lower); p.213 (lower); p.215 (right; lower left); p.217 (3); p.218 (2); p.219 (3); p.220(3); p.221 (2); p.222 (lower left); p.223 (2); p.224 (3); p.225 (3); p.226 (3); p.227 (5); p.228; p.229 (3); p.230 (upper); p.234 (3); p.235 (3); p.240 (upper right); p.241 (upper and lower left); p.244 (2); p.245 (3); p.246 (upper right); p.248 (3); p.249 (4); p.250 (2); p.251 (2); p.252 (3); p.253 (3); p.255 (right); p.258 (2); p.259 (3); p.260 (3); p.261 (upper; lower); p.262 (upper; lower right); p.264 (3); p.265 (center; lower left); p.266 (upper); p.267 (upper; center); p.272 (upper; center); p.277; p.285; p.289 (upper); p.290; p.300; p.301 (upper); p.302 ; p.303; Back Cover
National Archives and Records Administration (NARA)
p.46 (center); p.48 (upper and lower); p.49 (3); p.56 (upper); p.64 (upper right); p.66 (lower right); p.69 (upper left); p.89 (lower right); p.95 (upper); p.98 (3); p.99 (lower); p.100 (upper, lower right); p.101 (upper); p.103 (lower left, right); p.105 (upper); p.107 (upper); p.116 (upper); p.293
NARA via M. Pierre Trichet
p.28 (2); p.34 (upper and center)

Joel Powell
p.8 (3); p.15 (lower); p.36 (lower); p.71 (lower); p.72 (upper); p.104 (upper left; right); p.120 (center); p.143 (upper left; right); p.254 (left); p.256; p.261 (center); p.262 (left); p.265 (upper); p.267 (lower); p.268 (4); p.269 (3); p.270 (3); p.271 (3); p.272 (lower); p.273 (3); p.274 (3); p.275 (3); p.276 (lower); p.313; p.315
Redstone Arsenal
p.29 (insert, processed by M. Mathyk); p.52 (left); p.96 (inset); p.108 (right)
San Diego Aerospace Museum (SDAM, via Art LeBrun)
p.11 (upper); p.30 (center); p.31 (upper; lower left); p.32 (2); p.76 (upper and lower right); p.77 (center; lower left); p.79 (3); p.80 (3); p.111 (lower left; right); p.112 (left; upper right); p.116 (center; lower right); p.117 (upper; lower); p.121 (upper); p.125 (upper; lower); p.126 (3); p.127 (upper); p.128 (upper; lower right); p.129 (3); p.130 (upper); p.132 (upper left, right; lower left); p.136 (upper, lower left; upper right); p.140 (upper; center left, right); p.141 (upper); p.146 ((3); p.148; p.150 (lower); p.151 (upper left; lower left); p.167 (2); p.170 (3); p.171 (lower right); p.172 (3); p.173 (3); p.174 (3); p.175 (3); p.176 (3); p.177 (3); p.178 (upper); p.184 (lower left); p.186 (upper); p.187 (right); p.189 (lower left; right); p.190 (3); p.191 (3); p.192 (3); p.193 (lower left) p.194 (3); p.195 (left; lower right); p.206 (3); p.207 (3); p.208 (3); p.209 (3); p.210 (3); p.216 (left; lower right); p.233 (lower); p.237 (2); p.238 (left; upper right); p.278; p.291; p.316 (upper)
Courtesy Glen Swanson
p.121 (USAF, lower); p.147 (NASA, lower right); p.211 (USAF)
United States Air Force (USAF)
p.13 (upper); p.15 (2); p.17 (2); p.25 (2); p.29 (lower); p.31 (right); p.33 (upper); p.34 (lower); p.38; p.40 (upper); p.41 (3); p.42 (3); p.43 (2); p.44 (3); p.45 (left; center); p.46 (upper; lower); p.47 (3); p.48 (center); p.50 (3); p.52 (center; upper); p.53 (2); p.54 (2); p.55 (2); p.56 (center; right); p.57 (upper; lower; center via J. Bascom-Pipp); p.58 (3); p.59 (3); p.60 (upper); p.61 (3); p.62 (3); p.63 (3); p.65 (2); p.66 (upper left, right; inset); p.69 (2); p.70 (2); p.71 (upper); p.72 (lower); p.73 (upper; lower right); p.74 (2); p.75 (3); p.77 (upper); p.78 (3); p.81 (2); p.82 (3); p.83 (3); p.84 (3); p.85 (2); p.86 (3); p.87 (3); p.88 (3); p.89 (upper left); p.90 (3); p.91 (3); p.92 (3); p.93 (2); p.94 (lower); p.99 (upper); p 100 (left); p.101 (lower); p.102 (3); p.103 (upper); p.105 (lower); p.106 (2); p.107 (lower); p.108 (upper); p.109 (upper); p.110 (lower); p.111 (upper); p.112 (lower right); p.113 (center); p.118 (right); p.120 (lower); p.122 (upper; center); p.124 (lower); p.127 (inset); p.128 (left); p.136 (lower right); p.137 (upper); p.139 (upper; lower left); p.145 (lower); p.147 (upper left); p.154 (left; lower right); p.155 (lower left); p.156 (upper); p.157 (4); p.158 (upper left; right); p.159 (2); p.160 (upper and lower right); p.161 (upper); p.162 (3); p.163 (2); p.164 (3); p.165 (upper); p.168 (lower right); p.169 (3); p.171 (upper left); p.197 (3); p.198 (3); p.199 (upper); p.200 (center); p.212 (upper left; right); p.213 (lower left); p.214 (3); p.215 (upper); p.231 (right); p.232 (3); p.233 (upper); p.238 (lower right); p.239 (2); p.240 (left; lower right); p.241 (right); p.246 (left); p.254 (upper; lower right); p.255 (upper); p.257 (2); p.276 (upper); p.286; p.287 (upper); p.288; p.292; p.296; p.297; p.304 (lower); pp. 305-312 (Cape Guided Tour booklet)
University of Central Florida, Florida Space Coast History Project, courtesy Dr. Lori Walters
p.9 (2); p.13 (lower); p.19; p.30 (upper; lower); p.33; p.76 (left); p.110 (center); p.113 (upper); p.115 (upper); p.139 (right); p.144 (left; upper right); p.266 (left); p.280; p.284; p.287 (upper); p.301 (lower); p.304 (upper)
U.S. Navy
p.67 (2); p.95 (lower)

INDEX

1st Space Launch Squadron 18
45th Space Wing 7, 22, 36-37, 101, 318
6555th Test Wing 36, 98

A
Able and Baker 29, 64
AC-43 135
AC-68 137
aerodynamic fairing 167, 207
Aerothermodynamic/elastic Structural Systems Environmental Tests 230
Air Force Eastern Test Range 22, 36
Air Force Missile Test Center 14, 36
Air Force salvage crews 4, 143
Air Launched Ballistic Missile 108
Alpha Draco 4, 103, 105, 108
ANNA 196, 199
Antares II 233-234
Antigua 10, 36
Apollo 4, 12, 20, 22, 121, 133-134, 177, 223-228, 233, 274, 295, 314, 317
Apollo 1 20, 22, 121, 133, 227, 274
Applications Vertical Tests 228
Aries 5, 256-257
Aries rocket 257
Army Ballistic Missile Agency 16, 52-53, 223
Ascension 10, 22, 36, 138, 233
Athena 5, 22, 27, 257-258
Atlantic Missile Range 16, 22, 74, 79-80, 108, 140
Atlas A 16, 76, 113, 146, 284, 301
Atlas B 77, 113
Atlas C 168
Atlas D 128, 151, 188, 233
Atlas E 78-80, 237
Atlas I 5, 251-253, 268
Atlas II 251, 253-254
Atlas IIAS 251, 255, 298
Atlas III 20, 206, 251, 258
Atlas V 7, 24, 26, 36, 244, 258-263, 316
Atlas Vega program 206
Atlas-Able 5, 125, 127-128, 168-171
Atlas-Able Static Firing Disaster 125, 128
Atlas-Agena A 172-173
Atlas-Agena B 171
Atlas-Agena D 171-172
Atlas-Antares 5, 233
Atlas-Centaur 5, 7, 20, 27, 130-132, 135-136, 171, 176, 206, 209-210, 247, 291
Atlas-Centaur 43 136
Atlas-SCORE 5, 167
Azusa antennas 140

B
B-47 105-106
B-52 107-108
B-57 4, 106
B-58 105
Banana River 10, 24, 216-217, 266, 287
Banana River Naval Air Station 10
Bibby, Cecilia 187
Big Shot 5, 228-230
Birdwatchers 266
Black Goat 106
Blockhouse 11-12, 14, 16, 18, 20, 24, 30, 80, 89, 113, 188, 192, 209-210, 221-222, 225, 235, 255, 262, 270-271
Blue Scout I 4, 27, 99-101, 257, 270
Blue Scout II 99-100
Blue Scout Junior 97, 99
Bold Orion 4, 105-106, 108

Bomarc 4, 11-12, 27, 40, 45-46, 50, 95, 106-107, 276
Bull Goose 4, 16, 42, 46
Bumper 4, 10-12, 14, 16, 26-28, 40, 268

C
C-133 91, 121, 190
Cape Canaveral lighthouse 13
Cassini space probe 153, 244, 246, 299
Castor 4A 255
Castor solid rocket motor 204, 256
Centaur upper stage 24, 244, 259-260
Chaffee, Roger 20, 133
Chevaline 4, 68
Chimpanzee Enos 193
Chimpanzee Ham 178
Cocoa Beach 14, 22, 147, 184
combustion instability 128-129, 314
Commercial Titan 5, 239-241
Complimentary Expendable Launch Vehicle 24
Cooper, Gordon 112, 183-184, 187, 189, 195-196, 267, 290, 316-317
Courier 1A 197

D
Debus, Dr Kurt 16, 188
Delta 5, 7, 12, 16-18, 24, 26-27, 35, 46, 109-110, 114, 116, 121, 135-136, 138, 147, 151, 153, 163, 199-205, 230, 232-233, 244, 247-251, 256, 258, 263-269, 274, 276, 300, 303, 313
Delta 241 18, 138
Delta II 5, 7, 12, 16, 18, 26, 114, 153, 199, 247-251, 256, 263, 266-267, 269, 313
Delta III 5, 114, 256, 263
Delta IV 5, 7, 24, 26-27, 35, 121, 244, 258, 263-266, 268, 274, 303
Douglas S-IV 223
Dyna-Soar 24, 230

E
Eastern Space and Missile Center 22, 36, 318
Echo balloon 199, 201, 228-229
Eisenhower, President Dwight 153, 167, 183-184
erector gantry 16-17, 47-48, 81-82, 84, 93, 112-113, 218-220, 222, 274, 282, 288
Evolved Expendable Launch Vehicle 24, 251, 263
Explorer 6 105, 163, 166

F
Florida Space Authority 101, 256-257

G
Gemini 5, 12, 27, 32, 73, 89, 113, 118-119, 133, 141, 145, 177-178, 218-222, 266-267, 273-275, 292, 294, 301
Gemini-Titan 5, 218-220, 273, 292, 294
Glenn, John 174, 178, 180, 184-185, 187, 193, 195, 272, 289
Grand Bahamas 10, 22
Grand Turk 10, 22, 138
Grissom, Virgil "Gus" 16, 133-134, 142-143, 178, 182-183, 273, 316

H
Hangar S 16, 115, 118, 160, 183, 185, 289, 317
Hermes RV-A-10 4
High Virgo 4, 105, 108
Honest John transport vehicle 103
Horizontal Integration Building 265
Horizontal Integration Facility 24, 264
Hotrod series 77
Hound Dog 4, 107

Huntsville, Alabama 16, 50, 60, 223
Hyper Environmental Test System - 609A 97

I
ICBM Row 18, 35, 81, 83, 268, 278
Industrial Area 4, 12, 14-16, 19, 21, 33-34, 115-116, 278, 280, 287, 315, 317
Integrate, Transfer and Launch 25, 211-212, 293
Interim Upper Stage 240, 244
Inter-Pad Missile 314
interstage adapter 173, 176, 206-207

J
Jason 4, 11, 95, 102
Joint Long Range Proving Ground 10, 36-37, 40, 71
Jonathan Dickenson Missile Tracking Annex 22
Joust 5, 256
Juan Ponce de Leon 10
Juno I 5, 12, 27, 96, 117, 153-156, 285
Juno II 5, 12, 27, 127, 147, 153, 156-158, 304
Jupiter A 4, 50, 52, 314
Jupiter C 4
Jupiter IRBM 4, 16, 52, 60, 96, 314

K
Kennedy Space Center 16, 22, 24, 51, 118, 143, 145, 150, 152-153, 215, 266, 271
Kennedy, President John F. 22, 183
Kraft, Christopher C. 118
Kranz, Gene 118

L
Lark 4, 12, 27, 40, 280
Launch Area Support Ship (LASS) 71
Launch Complex 11 - 30-31, 80
Launch Complex 12 - 128-129
Launch Complex 13 - 237, 315
Launch Complex 14 - 171, 177, 184, 188-189
Launch Complex 15 - 93
Launch Complex 16 - 54
Launch Complex 17 - 18, 57, 268
Launch Complex 18 - 29
Launch Complex 19 - 32, 113, 218
Launch Complex 20 - 17, 256
Launch Complex 21 - 41-42, 46
Launch Complex 26 - 275
Launch Complex 3 - 12
Launch Complex 34 - 223
Launch Complex 36 - 23, 26
Launch Complex 37 - 22
Launch Complex 41 - 259, 302
Launch Complex 46 - 22, 257
Launch Complex 5/6 - 14, 279
Launch Complex 9 - 16
LC-11 - 4-5, 77
LC-12 - 5, 31, 74-75, 79, 168, 170, 173-177
LC-13 - 74, 77, 79, 171-172, 175-177, 237, 264
LC-14 - 5, 20, 75-76, 78, 172-173, 189-191, 194, 196, 267
LC-15 - 4, 18, 81-83, 89, 92
LC-16 - 4, 18, 53-54, 82-83, 91
LC-17 - 4-5, 18, 26, 57, 164, 200, 266
LC-19 - 5, 18, 83, 89, 218
LC-20 - 5, 18, 81, 84-85, 256
LC-21 - 42, 46
LC-25 - 4
LC-26 - 4-5, 157

LC-29 - 4, 18
LC-30 - 4, 20
LC-31 - 4, 20
LC-34 - 5, 20, 22, 188, 223
LC-36 - 5, 206, 210, 251, 264
LC-37 - 20, 22, 188, 227, 303
LC-39 - 225, 266
LC-40 - 5, 24-25
LC-41 - 5, 24, 217
LC-43 - 4, 101
LC-46 - 4-5, 22, 257
LC-47 - 4, 101
LC-5 - 155, 180-181
Lewis, Jim 142-143
Liberty Bell 7 4, 142-143, 183, 187
Lockheed Martin's "clean pad" 302
Loki-Dart 102
Long Cable Mast 108
Lunar Orbiter 165, 168, 172, 177, 237
Lunar Prospector 257-258
M
Mace 4, 27, 41-42, 276
Mackey, Skip 255
Malmstrom AFB 85
Mariner 67, 117, 147, 150, 171, 175-177, 237
Mark 3 78, 151
Mark 4 79-81, 83, 85, 91
Mark 6 89-91
Mars Observer 239-241
Mars Pathfinder 152
Mars Reconnaissance Orbiter (MRO) 259
Mars Rover Spirit 153, 300
Matador 4, 14, 16, 36, 41
Medaris, General 16, 314
Mercury 5, 12, 16, 27, 118-119, 129-130, 142-144, 150, 177-180, 182-196, 206, 266-267, 269, 272-273, 275, 290, 316-317
Mercury-Atlas 5, 27, 130, 188-194, 267, 290, 316
Mercury-Atlas 3 abort 130
Mercury-Redstone 5, 12, 178-180, 183, 186
Merritt Island 301
Minuteman I 4, 27, 85, 87, 90, 121
Minuteman II 85, 88-89
Minuteman III 20, 22, 68, 85, 89
N
National Reconnaissance Office 255
Navaho 4, 12, 14, 16, 20, 27-29, 34, 47-49, 95, 103, 105, 107, 112, 120, 122-123, 161, 270-271, 276, 282, 287
Naval Ordnance Test Unit (NOTU) 71-72
Naval Research Laboratory 159
Nike Cajun sounding rocket 102
Nike Smoke 102
Nike Tomahawk 102
O
O'Hara, Dee 185
Orbital Test Satellite 135
Orbital Test Satellite (OTS) 135
Orbiting Astronomical Observatory 176
P
Pad 19 white room 221
Pan American Airways 14, 32, 145
Patrick Air Force Base 10, 12-15, 36, 105-106, 317-318
Pegasus 14, 22, 224-225
Pershing I 4, 27, 33, 53-54
Pershing II 4, 54-55

Photographic Services at Cape Canaveral 4, 145
Pioneer 12, 150, 158, 163, 165, 168, 183, 266, 318
Pluto New Horizons spacecraft 260
Pogo vibrations 89
Polaris A-1 4-5, 65, 236
Polaris A-2 67
Polaris A-3 65-68, 72-73, 108
Polaris FTV 4, 27, 65
Polaris-STAFF 236
Port Canaveral 8, 14, 67, 71-72, 121, 139, 142-144
Poseidon 4, 12, 18, 22, 27, 67-69, 74, 138
Powers, Col. John "Shorty" 118-119
Press Site 1 138, 266-268
Press Site No. 2 267
Project Argus 95
Project FIRE 233-235
Project Hermes 93
Project SCORE 167
Project STAFF 236
Prospector rocket 256
Q
QB-17 drone targets 106
R
Range Safety Officer 15, 115, 121, 128, 136, 175, 314-315
Ranger 12, 116, 150, 165, 171, 173-174, 315
RCA Corp. 145
RD-180 251, 258
Reagan, President Ronald 247, 251
Red Tigress program 256-257
Redstone 4-5, 11-12, 14, 16, 27-29, 50-53, 60, 96-97, 108, 118, 142, 153-156, 178-180, 182-183, 186, 223, 269, 273, 314, 319
Redstone Arsenal 16, 50, 60, 319
Research in Supersonic Environment (RISE) 47
RL-10 engines 224
Robin weather rocket 101
Rogers, Chuck 145
RS-68 hydrogen-oxygen rocket engine 263
Ruins of Cape Canaveral 5, 268
S
Saturn I 20, 22-23, 35, 188, 223-224, 226
Saturn IB 20, 22, 27, 134, 223, 226-228, 295, 314
Schirra, Wally 133, 220
Shepard, Alan 16, 178, 180-181, 186, 273, 316
Ship Motion Simulator 18, 33
Signal Communications by Orbiting Relay Equipment 167
single-engine Centaur 260
S-IVB upper stage 223, 226-227
Skid Strip 4, 14, 19, 28, 43, 107, 112, 116, 120, 168, 190, 315
Skybolt 4, 68, 103, 108
Snark 4, 12, 14, 27, 43-45, 120, 138, 276, 315
Solid Motor Assembly 24-25, 212
Space and Missile Museum 18, 276
Space shuttle Challenger 272
Spin Balance Facility 131
Stafford, General Thomas 133, 220
Starbird rocket 256
Starlab program 256
Static test firing 47, 168, 170, 190

Stellar Acquisition Flight Feasibility 236
Stellar Inertial Guidance System 236
Strategic Air Command 43, 92
Super Loki 101
T
tail service mast 20, 224
Technical Laboratory 14-15
Telemetry mast 72, 74
Thor 15-16, 18, 22, 55-60, 94, 101, 109, 114, 116, 122, 163-164, 171-172, 196-201, 228, 230, 232, 247, 270, 281, 283, 314
Thor tactical pad 59
Thor-Able 109, 163-166, 199
Thor-Able-Star 5, 196-199
Thor-Delta 5, 17, 109-110, 116, 121, 151, 163, 199-201, 230, 232-233, 269
Tiros 1 163, 166
Titan 34D 5, 24, 211, 239-241, 316
Titan I 4, 12, 17-18, 27, 33, 79, 81-85, 89, 113, 128, 140, 218, 256
Titan II 4, 9, 18, 27, 89-93, 145, 211, 215, 218-220, 267, 288, 292, 314
Titan IIIA 5, 18, 27, 211, 256
Titan IIIC 5, 18, 24-25, 27, 211-217, 239, 241, 244, 293, 315
Titan IIIE Centaur launch vehicle 216
Titan IVA 24, 244, 246, 316
Titan IVB 24, 153, 244, 246-247, 299
Titan mission patches 316
Transit 3A 198
Transtage 211, 213, 241
Trident I 4, 18, 22, 27, 68-70, 136
Trident II 4, 22, 27, 68-72, 74, 101, 137, 257, 296
U
Undertakers 143
USS Norton Sound 95
USS Observation Island 18, 67
V
V-2 4, 10, 40, 50, 268
Vandenberg Air Force Base 57, 78, 128, 171-172, 196, 199, 211, 228, 316
Vanguard 5, 12, 16, 20, 27, 47, 98, 105, 115, 124, 156, 159-163, 269, 286
Vela nuclear detection payload 237
Vertical Integration Building (VIB) 212
Vertical Integration Facility (VIF) 26, 36, 258-259, 261
Vertical Launch Facility 26 16, 29
Viking 1 Mars probe 152
von Braun, Wernher 20, 50, 60, 188, 223
Voyager spacecraft 152
W
WAC-Corporal 12, 40
Wendt, Guenter 143, 186, 318
White Sands Missile Range 10
White, Edward 133, 185
X
X-10 4, 14, 107, 120
X-17 4, 65, 94-95
X-20 24, 230
XM-20 Sergeant motor 94